T0251444

DESIGN METHODOLOGY IN ROCK ENGINEERING

Creative design and innovative construction resulted in this spectacular excavation: the UK Crossover Section in the Channel Tunnel between England and France (photograph by QA Photos Ltd, courtesy of TML, the Channel Tunnel Contractors, London).

DESIGN METHODOLOGY IN ROCK ENGINEERING

Theory, education and practice

Z.T. BIENIAWSKI

Professor of Mineral Engineering
The Pennsylvania State University, University Park

A.A.BALKEMA / ROTTERDAM / BROOKFIELD /1992

CIP-DATA KONINKLIJKE BIBLIOTHEEK, DEN HAAG

Bieniawski, Z.T.

Design methodology in rock engineering: theory, education and practice / Z.T. Bieniawski.
Rotterdam [etc.]: Balkema. – Ill.
With bibliogr., index.
ISBN 90 5410 126 1 bound
ISBN 90 5410 121 0 paper
Subject heading: rock engineering.

Published by
A.A. Balkema, P.O. Box 1675, 3000 BR Rotterdam, Netherlands
A.A. Balkema Publishers, Old Post Road, Brookfield, VT 05036, USA

ISBN 90 5410 126 1 hardbound edition
ISBN 90 5410 121 0 student paper edition

Contents

Preface

Design is to the '90s what finance was to the '80s and marketing to the '70s; it is the corporate buzzword for the new decade.

Business Week (September 4, 1990)

Design: the very word epitomizes creativity, innovation, and the essence of engineering. After all, design and engineering are inseparable: engineers design!

Derived from the Latin word *designare* – to map out – design may be formally defined as that socio-economic activity by which scientific, engineering and behavioral principles, allied with technical information and experience are applied with skill, imagination and judgment in the creation of functional, economical, aesthetically pleasing, and environmentally acceptable devices, processes, or systems for the benefit of society (Gregory, 1986).

This is a lengthy definition but it conveys an important message that, in essence, a good designer needs not only knowledge FOR designing (technical knowledge that is used to generate alternative design solutions and select the best among them) but also a good designer must have knowledge ABOUT designing (an appropriate process or systematic framework to follow). In fact, a good product can only result from a suitable combination of good designing, good technical knowledge, good goals, good tools, good analyses, good designers, and good management (Hubka, 1987).

Design is one of the oldest human endeavors going back to prehistoric times when mankind conceived hunting implements, shelters and clothing. These and later inventions preceded the development of the sciences by many centuries. However, to this day design is being done intuitively as an art rather than a science. Moreover, it is one of the few technical areas where experience is more important than formal education (Suh, 1990). As a result, there is an ad hoc approach both to design and to design education.

While engineers can point to amazing achievements throughout history, in some cases the ad-hoc decision-making process has not been effective, leading to serious failures both structural and economic.

In the preceding prosperous years, the mineral industry in America has made spectacular technological advances, as well as profits, arising from favorable mineral exploration conditions, as well as accumulated experience and sheer entrepreneurship. Somewhat suprisingly, design innovation *per se* played little part in this success story because the mineral industry traditionally does not frequently introduce drastic design changes, unlike the automobile and aerospace industries. For example, in the mining industry, design considerations consist of a gradual change in certain excavation features or the addition of new rules of thumb rather than the utilization of a systematic process of engineering design aimed at boosting creativity and innovation. This does not mean that mining engineers are any less capable but somehow much of the technological innovation in mining takes place without a deliberate adherence to the design process.

Recently, a new discipline has emerged: design theory and methodology, prompted by a special study of the American Society of Mechanical Engineers (Rabins, et al. 1986). It resulted in the establishment of the Design Engineering Division within the ASME which passed a strong message to other fields that design methodology is crucial to the success of a new product, process, or construction project.

The field of rock engineering has an unsurpassed potential for design innovation. *Rock engineering* signifies that broad range of disciplines which contribute to the science and engineering of rock materials and rock masses, and is a field of professional practice and research which draws heavily on the elements of rock mechanics. Other disciplines that contribute to rock engineering include civil engineering, mining engineering, geological engineering, petroleum and natural gas engineering, mechanical engineering, and engineering geology (including elements of geohydrology and geophysics). These disciplines are involved with the design and construction of such projects as mines, tunnels, foundations for structures, excavated rock slopes, dams, shafts, boreholes, oil reservoir wells, and underground storage facilities for oil, gas and nuclear waste. Rock engineering allows us to improve our environment, mitigate natural hazards, and construct engineered facilities to improve our quality of life.

In the field of rock engineering, only limited attention has been paid to the design process (Bieniawski, 1984). The purpose of this book is to introduce the principles of engineering design theory and methodology and identify the main stages of the engineering design process. A design theory and methodology specifically for rock engineering is proposed and its practical use demonstrated on the basis of important case histories. From problem formulation, through analysis and synthesis, to evaluation and optimization, the design process is seen as the use of engineering heuristics in a systematic framework leading to innovative problem-solving.

To preserve the momentum of the design message, design education is also discussed and suggestions are presented for improvements in the training of future design engineers.

In essence, the logic behind the development of the topics in this book is to start with an appraisal of current trends concerning global design activities and competitiveness (Chapter 1) and to introduce the concepts of creativity and innovation as being inseparable from any design (Chapter 2). An insight into how designers design is also included in Chapter 2. The state of the art in engineering design is then discussed in Chapter 3 with a detailed exposé of all significant design theories and methodologies. This leads directly to the main thrust of the book which is the issue of design in rock engineering. This subject is treated in three chapters appropriate to a modern discipline: theory (Chapter 4), education (Chapter 5), and practice (Chapter 6). The last chapter, Chapter 7, is devoted to skills development, presenting the designer with an extensive repertoire of widely available tools and concepts. The Appendix lists a compendium of useful design charts for rock engineering, traced after a thorough literature search. A Bibliography concludes the book with an up-to-date list of references.

The ideas presented in this book were developed and compiled by the author over many years, as a researcher, educator, and practical designer, but the actual writing was only possible in the academic year 1990/91 during the author's tenure at Harvard University as Visiting Professor in the Graduate School of Design. The co-operation and assistance of Harvard University and, in particular, of Prof. William J. Mitchell, is gratefully acknowledged. My research assistant at Penn State, Dr Dwayne C. Kicker, made a valuable contribution by searching the early literature on engineering design and reviewing a design case history in coal mining.

My wife Elizabeth – mobilizing her graduate research skills in library science and her expertise in garden design – was most helpful in editing and cross-referencing the text and the index, while my son Stefan – a graduate student in aerospace engineering – reviewed the manuscript and provided a refreshing perspective as a future design engineer.

University Park, Pennsylvania   Z.T. Bieniawski
July 1991

CHAPTER 1

Introduction

The proper study of mankind is the science of design.

Herbert Simon (1969)

Why are so many managers and engineers mystified by design? Why are we not supported by the firm foundation of a science of design? Is it because our understanding of design, including engineering design, is far from mature?

To answer these questions, one should first note that to be a modern discipline, it must consist of three elements, *theory, education and practice*, to guarantee its capacity for renewal and development. In the case of design, these three elements have not received sufficient attention and – until recently – have not been a subject of concentrated effort on the part of researchers, educators, and practitioners.

Indeed, there is currently a realization that design has been underutilized and misunderstood. The meteoric rise of post-war Germany and Japan coupled with the outstanding performance of some American companies like IBM, DuPont, General Electric, Apple Computer, and Procter & Gamble, is testimony to the power of good design.

But the greatest recent impetus to the importance of design came from the realization that since everything has to be designed, design needed leadership as did many other business functions. 'Design is a central activity to line managers and it contributes directly to company profitability,' was a graduation message to the new generation of business leaders at a prominent Business School (Gregory, 1987). It continued, 'To be effective, the design message must be understood within the company at all levels, from chairman down, and not merely as a fashionable artifact.'

Yet, where will this design understanding come from? It can come from training to create an awareness of design and to determine what the company's needs are and how they can be met. And this is best achieved by understanding design theory, by engaging in design education, and by reviewing good design practice. In this way, the engineering design process will not only assume its rightful role, but will also pay handsome dividends in ever increasing market competition.

Lest a sceptical reader dismiss this argument, the importance of design was underlined by the *Time* magazine's *Designs of the Decade* announced in 1990 along with its well-known Man-of-the-Year award. Among the first winners were:

Apple Macintosh Computer, Mazda Miata, Battery Park City and the Vietnam Veterans Memorial.

A growing concern that we must elevate the importance of design in our engineering practice as well as in the curricula of engineering universities is further manifested in a series of annual international conferences on engineering design, in the publication of specialized journals, and the establishment of a National Science Foundation program specifically devoted to supporting research in design theory and methodology. There are economic reasons which make engineering design important stemming from the need for nations to compete globally but design is even more fundamental to mankind than economic success. Design is important because it determines the ultimate outcome of engineering activities, an improvement in the quality of life and in the provision of energy needs. Yet, while we may appreciate design as an outcome of a process, we do not understand the process that produces the outcome and cannot quantify evaluation of a design. Clearly, to obtain better performance, one requires correct principles and methodologies to guide decision-making in design. There are good design solutions and unacceptable design solutions, so that there must be features to distinguish between good and bad designs (Suh, 1990).

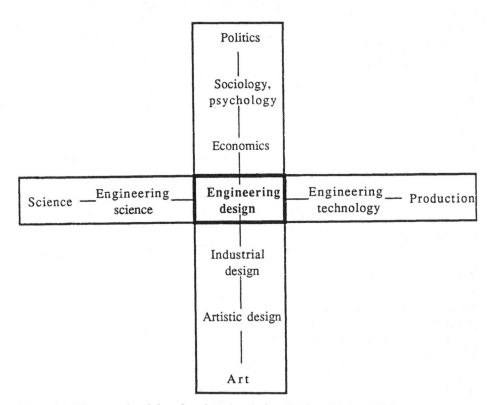

Figure 1.1. The central activity of engineering design (Pahl and Beitz, 1984).

1.1 HISTORICAL PERSPECTIVE

For many generations, designers expressed the need for a better understanding of design. In fact, one of the simplest criteria to judge a design is also the oldest statement on the subject. Around 30 B.C. the Roman Architect, Vitruvius, listed these three essential qualities of good design: '*commodity, firmness, and delight.*' These words are from an old Latin translation but in modern terms *commodity* means usefulness and fulfilling its function well; *firmness* means being well-built and long-lasting, while *delight* implies character and beauty.

Engineering design has been historically viewed as a form of art and not as a technical activity. It has been thought that design primarily involves creativity and intuition, which are the spontaneous skills of the designer. Therefore engineering design was viewed as being spontaneous and experience-dominated. Intuitive design has also seemed to be an opposing idea to systematic design.

Yet, through the years, concepts of systematic design have begun to emerge. An excellent historical background and discussion of the development of systematic design is provided by Pahl and Beitz (1984). Its origins may, perhaps, be traced back to the great designer, Leonardo da Vinci. While thoughts of systematic design have been around for a long time, one of the charter works attributed with the revival of modern design methodology was a paper by Asimow (1962). Begining with a discussion on the philosophy of design, he submitted that the principles which lead to design are based on one's own experiences. Choices of principles will thus vary from person to person, and hence more than one philosophy of design will exist. Asimow identified three components of a design philosophy: (1) a set of principles; (2) a framework which leads to action; and (3) a critical feedback which evaluates the design.

Another important contribution was by Jones (1963) at the Conference on Design Methods (Jones and Thornley, 1963). He presented a method of systematic design which outlined an organized notation of all design information with guidelines on the effective use of this information. The three main stages of the design process were identified as analysis, synthesis, and evaluation. Several sub-categories which described the process for each stage were also provided. By outlining the design process in this detail, the logical analysis could be thoroughly recorded.

Glegg (1969) has attempted to offer some practical guidelines for design that could be useful in industry, suggesting that a clear definition of the problem was essential. He focused on the characteristics of the creative designer and identified three personal traits: the inventive, the artistic, and the rational. Suggestions for inventiveness included the importance of concentration and relaxation and the avoidance of the influence of tradition. Also, he noted that expert knowledge in a field was not necessary for inventiveness. Many inventions were produced by people who were not formally trained in the field of their invention.

During the 1970's, interest in the design methodology movement was spurred on by the Design Methods Group, which is the US counterpart to the Design Research Society of Great Britain. These organizations co-sponsored a conference in 1973 in London, drawing over 400 delegates. However, no formal proceedings were published. The next major conference was held in Portsmouth (Evans et al. 1982). The proceedings were structured to review engineering design in the past, to discuss the present status of design, and to project the nature of design for the future.

Given all the design methods research in those years, Fogue (1982) realized the importance of producing a practical design theory which could be implemented in practice. He suggested that creativity simply involved optimizing the structured design activities.

One of the early works on design methodology was presented by Archer (1984) who developed a procedural checklist for design consisting of eight stages which were sub-divided for a total of 229 sub-stages! The primary steps in this process included programming, data collection, analysis, synthesis, development, and communication. Archer emphasized that the act of designing involves a 'creative leap' from studying the problem to determining a solution.

The importance of a heuristical approach in problem solving and in design has been recognized by Koen (1984) who applied these concepts to the engineering method, i.e. to engineering design. Because of the importance of Koen's work, detailed discussion on this topic is included in Chapter 3.

Engineering problems are typically ill-defined. Rittel and Weber (1984) discussed the idea of such problems, also called wicked problems, which have no clarifying characteristics and consequently no single solution. Therefore, wicked problems do not yield 'true' or 'false' solutions, but 'better' or 'worse' solutions (Rittel, 1984).

It was suggested by Eder (1987) that the majority of design research has been directed toward the knowledge that is used by designers and little attention has been given to knowledge about the nature of the designer's work. He proposed that the study of design should cover these aspects: (1) the designers (their characteristics, working methods, etc.), (2) the activity (procedural aspects, creativity factors), (3) the object to be designed, (4) the context in which engineering design takes place, and (5) the social, moral, and political context of the use of the resulting technical system. Eder enounciated a comprehensive theory of design proposed by Hubka (1987): the theory of technical systems. This is treated in detail in Chapter 3.

Perhaps the most important modern contribution to the study of design theory and methodology is a series of bi-annual conferences entitled, *International Conferences on Engineering Design* (ICED 81 through 91). ICED 87 was the first of this series to be held in America, and was jointly sponsored by the American Society of Mechanical Engineers (ASME) and the Workshop Design-Konstruktion (WDK) which is an international society of design scientists.

According to Stauffer, Ullman, and Dietterich (1987), there is an urgent need for a better understanding of the design process because:

1. There is no objective way to evaluate the design process;

2. There is little understanding of how the design process can be improved;

3. An understanding of design methodology is essential for future development of computer-aided design (CAD) tools.

Designers base many decisions on qualitative rather than quantitative reasoning. Designers often find satisfactory rather than optimal results. To choose the best solution from among many solutions is one of the designer's greatest problems. There is a need to direct more attention to the solution of 'open-ended' problems (problems with many solutions and many routes to solutions). To this end, a sound design methodology is vital (Gill, 1987).

In the field of rock engineering, very little attention has been paid to design methodology. The only textbook specifically on this subject (Bieniawski, 1984) introduced the topic to the mining and tunneling community and pointed out the need for research in this area. However, the mining and civil engineering industries have as yet to support any research endeavor in design methodology.

Most recently, a significant contribution to design theory and methodology was provided by Suh (1990). He proposed the concept of an axiomatic approach to design which stipulates that there exists a fundamental set of principles that determine good design practice. These principles involve determination of func-

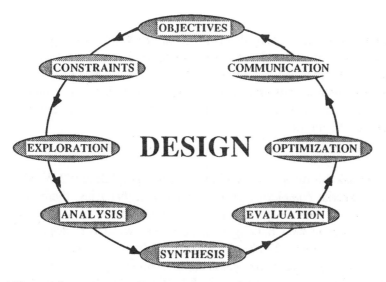

Figure 1.2. A representation of the design process as it appeared on the front cover of *Rock Mechanics Design in Mining and Tunneling* (Bieniawski, 1984). The counterclockwise direction of the arrows signifies a challenge to design engineers who have to move 'against the tide' of prejudice which views design studies as a 'soft' subject for engineers!

tional requirements and design parameters satisfying two axioms : the independence axiom and the information axiom. Suh (1990) also points out that case studies are indispensible to the learning process since they provide opportunities to examine the design axioms from many different points of view. Suh's work is of particular significance and is presented in detail in Chapter 3.

A number of lessons can be learned from this historical review of design activitities.

Lesson 1

The history of design is full of examples of individuals who invented things quite outside their own fields. For example, the first corn-reaper was invented by an actor who demonstrated it on stage. The first practical submarine was invented by an Irish schoolmaster in New York who wanted to sink the British navy!

The first calculating machine was invented by a 19-year old boy to help his father who was a wages clerk. While still at school, an American schoolboy developed his own electronic TV system which was so important that big corporations had to spend more than $2 million trying to overcome his patents. The domestic vacuum cleaner was invented by a builder of huge bridges, the universal joint by an astronomer, and the hovercraft by a radio engineer.

The lightning conductor was invented by Ben Franklin – United States Ambassador to France. The tank was invented by that most unlikely of all groups: the British Naval Air Service. The razor blade was invented by a 40-year old salesman (King Camp Gillette) who was determined to invent *something*.

Lesson 2

Nevertheless, most inventions were by determined entrepreneurs working in their fields, often on the basis of trial-and-error but sometimes discarding an idea too soon. For example, Thomas Edison invented the light bulb in 1879, but he stubbornly clung to direct-current motors, rejecting the invention of his onetime assistant, Nicola Tesla. In 1888, this Serbian immigrant patented his alternating-current 'electromagnetic motor.'

Lesson 3

One should always ask this question: 'Is this invention a good design'? Would it satisfy Vitruvius' criteria: *commodity, firmness, and delight*? Alas, there is no shortage of examples of poor design! As reported by the Presidential Commission, the cause of failure of the Challenger Space Shuttle was faulty design of a pressure seal. The $1.5 billion Hubble Telescope is also an example of poor design; a team of 40 people spent four years and $70 million to wind up with a near-sighted space telescope. According to NASA, 'the mirrors have been made very precisely but designed very precisely wrong.' Another example is the new Soccer Stadium in Milan, Italy, where it was overlooked that the mandatory grass surface was not possible under a closed roof!

Lesson 4

It is not surprising that when a good design is evolved it has to be guarded closely. IBM is faced with the nightmare of fighting imitations of their computers which are even promoted as 'IBM compatibles.' According to *USA Today* (June 15, 1990), Honeywell USA alleges that the Minolta company of Japan stole its patented autofocus technology, now featured in the Minolta Maxxum camera introduced in 1985. Apparently, in 1979, Honeywell invited a group of Japanese to visit its laboratories and demonstrated to them the autofocus technology, after which – claims Honeywell – Minolta agreed to use the information to design cameras with an autofocus system made by Honeywell. Instead, Minolta allegedly copied Honeywell's designs and never paid any royalties (*USA Today*, June 15, 1990). Minolta denied these allegations.

Lesson 5

The design process cannot be carried out efficiently if it is left entirely to chance, so a more systematic approach is essential. The aim of a systematic approach is to make the design process more visible and comprehensible so that those providing inputs to the process can appreciate where their contributions fit in.

1.2 CURRENT TRENDS

The current design scene is best characterized by comparing activities in the leading countries: USA, Japan, Germany and Britain, and by considering three aspects: R&D expenditures, quality of human resources, and government policies.

Starting with the United States, the history of design in this country is rich in great achievements: from the telephone to radar and laser technology, from mass-produced cars to lunar travel, from mechanical computers to microprocessors, and from mechanized agriculture to robotics. But by the early 1970s, the United States was slipping and foreign products started to penetrate and dominate American markets with such items as: cameras, electronic equipment, automobiles, and computers. Engineering design in the USA was on the decline, compounded by the fact that it was rarely recognized as a legitimate field for university research. What were the reasons for this situation?

Some statistical data are of particular interest. In 1988, Japanese universities awarded 74,000 BS engineering degrees to USA's 67,000. However, six engineers graduated in Japan for every scientist. In the USA, 1.4 science majors graduated to every engineer. Most of all, 95% of Japanese students perform better on mathematics and science examinations than the top 5% of US students. Nevertheless, the USA still employs more R&D personnel (750,000 in 1988) than the combined R&D workforce for Japan, Germany, Britain, and France.

With reference to Table 1.1, Japanese companies place a much higher priority

Table 1.1. Business-funded R&D (1987) ($ billions).

Country	Amount	% of GNP	% of all R&D	All R&D
USA	58.2	1.42	49.8	119.67
Japan	42.9	2.14	77.8	55.14
Germany	14.9	1.69	58.8	25.34

Source: *Mechanical Engineering*, April 1988: 31.

Table 1.2. US Government R&D spending ($ millions).

Type of research	Civilian	Defence	1990 total	1991 prop.	Percentage
Basic research	10,459	939	11,398	12,366	17.4%
Applied R & D	13,375	38,938	52,313	55,773	78.3%
Facilities			3,023	3,059	4.3%
Total	23,834 (37.4%)	39,877 (62.6%)	66,734	71,199	100.0%

Source: *The budget of the United States for Fiscal Year 1991*, Washington, D.C.

on R&D than their American counterparts. Although roughly similar amounts are spent in both countries on business-funded R&D, in Japan this constitutes 77.8% of that nation's total research effort. The USA figure is 49.8% and the US government's R&D involves 62.6% for defense research (see Table 1.2). Note that R&D yields far fewer benefits to the civilian economy than was once expected (Alic, 1988).

The Japanese government sponsors programs to assist with research technology transfer to industry in order to speed its application, cut training costs, and develop technically advanced products of commercial significance, e.g. Japan's fifth generation computer project and related software. Moreover, the multinational corporations of Japan have a good deal of control over their technology; nations with open economies do not. In addition, the hourly wage in the United States is about twice as high as it is in Japan and about nine times as high as the average in Korea.

The latest trend in the USA is not encouraging. According to *Business Week* Innovation Issue, September 1990, the total US industry spending on R&D in 1989 was $65.2 billion, with the downward drift in R&D as percentage of sales continuing. However, some American companies, e.g. IBM, Cray, Hewlett-Packard, have boosted their R&D spending sharply. On the negative side, only eight US companies in the mining business spent more than $1 million or 1% of their sales on R&D. Overall, the US mining companies spent on R&D only 0.22% of the value of production ($10.4 billion), as reported by NRC (1990). The report concluded: 'The mining industry is in a state of R&D stagnation. Most of the technologies currently in use were developed at least 20 years ago.'

Nevertheless, the current situation is that the United States spends twice as much on research as do Japan and Germany combined but this is primarily basic research, the findings of which are freely available. This is not surprising considering that this country has over 3000 universities and colleges most of which are active in research to some extent. Yet, while the United States is the clear leader in basic research and creative idea generation, it does not capitalize on these ideas, and when it comes to technological innovation and design, it is overtaken by its foreign rivals, particularly the Japanese. They are far more effective in using the design process as a vehicle for turning a creative idea into an innovative product. This overrides the fact that the USA has the largest pool of trained scientists/engineers in the free world, some 4.5 million.

1.3 EMERGENCE OF A NEW DISCIPLINE: DESIGN ENGINEERING

Design research in the leading industrial countries can be traced back for a number of decades but some other countries, such as Poland, have also made significant contributions in this area.

1.3.1 *Japan*

The most prominent leader in the field is Hiroyuki Yoshikawa of the University of Tokyo, Department of Precision Machinery Engineering who started to work on *sekkeiron* (design theory) around 1970. However, his ideas have not been accepted nor understood by the research community until 1985 when he organized an IFIP (International Federation of Information Processing) conference on Design Theory for CAD in Tokyo.

Today, the importance of engineering design study is widely recognized in Japan and H. Yoshikawa, now assisted by Tetsuo Tomiyama, remains the acknowledged prime mover. The Japanese engineering design community attracted inter-disciplinary interest from many fields including mechanical engineering, computer science, artificial intelligence and others. Three reasons were given for this:

1. The introduction of CAD/CAM technology (computer-aided design/computer-aided manufacturing) was expected to reduce the designer's work in manual drafting and provide more time for the creative aspects of design. Apparently CAD/CAM did reduce drafting time but the latter did not happen. Thus, it is more crucial to understand design processes than to emphasize better representation of design objects.

2. Artificial intelligence researchers are getting more interested in design, primarily due to the creative nature of design which is considered typical of

human intelligence, and secondly, due to the practical industrial applicability of the subject.

3. The scale and complexity of designed objects has increased dramatically in the last two decades due to advances in technology, particularly in micro-electronics. Increasing competition in industry mandates that the design be performed in a shorter time, at less cost, and with higher quality. Engineering design practices must be drastically revised to meet such demands.

According to Tomiyama (1990), early work in engineering design in Japan was influenced by the engineering science movement in the USA in the 1960s. Subsequently the Taguchi method was proposed as a method for quality control but it has not been recognized as a design methodology. CAM/CAD technology was, however, considered a critical tool for better productivity.

As said earlier, Yoshikawa was most influential in the history of Japanese engineering design research. He insisted that the central emphasis of engineering should be design and proposed that more stress should be placed on synthesis than analysis. This was quite contradictory to the most influential ideas of the time, i.e. engineering science. He was also interested in what governs or regulates design processes. In this context, the General Theory of Design was developed (Yoshikawa, 1981) aimed at explaining and understanding design processes generally regardless of the discipline being served.

The main idea of the General Theory of Design is that the power of design, as a creative intellectual activity, results from operations of concepts, from design objectives to design solutions, based on an axiomatic set theory which explains how the designers use their knowledge. This theory is discussed in Chapter 3.

Currently, engineering design research in Japan is promoted by three organizations: Japan Society of Precision Engineering as the biggest community for CAD/CAM research, the Japan Society of Mechanical Engineers which is planning to set up a division for engineering design (as was done by the ASME), and the Japanese Society for Artificial Intelligence which is interested in expert systems design and knowledge-based CAD systems.

Design research pursued at Japanese universities includes the University of Tokyo where Tomiyama and Yoshikawa are working on a computable model of design processes; at Waseda University Nakazawa is working on design methodology having introduced Suh's axiomatic design principles; and at Kyoto University a group under Okino is interested in the modeling of design objects using a technique called *molelon*.

According to Tomiyama, the overall aim of all these research activities is to proceed from a design theory to an implementable methodology which could not only be used as a guiding principle but would also provide practical instruction on how to design.

1.3.2 *Europe*

Major design research activities in Europe have been centered in Germany where the work of Pahl and Beitz (1984), *Engineering Design*, is the classic in the field and even led to the development of industrial standards for design methodology. However, other countries notably Britain, Switzerland and Poland have made important contributions and will be discussed first .

In Britain considerable attention is being given to improve the quality of engineering design (Wallace and Hales, 1987) and the systematic design approach of Pahl and Beitz is being prominently used at Cambridge University to improve the structure of design teaching and to contribute to design research.

For many years, the *Design Council* in Britain has campaigned for greater emphasis on the quality of engineering design and has been responsible for many important initiatives, both in education and in practice. The *Design Research Society* was formed in 1966 with strong influence from architecture but with weak participation from engineering. The *Institution of Engineering Designers* which goes back many years is still generally regarded as a 'junior' institution not of the caliber of the 'proper' learned societies like the British Institution of Mechanical Engineers. However, the excellent journal *Design Studies*, initiated in 1979 with a strong architectural bias, has in recent years given prominent coverage to engineering design topics and has grown to international stature. Another journal, *Engineering Education and Training*, includes advice and problems for teachers of engineering design.

An important contribution is now being made by a body known as SEED (Sharing Experiences and Engineering Design) initiated to act as a clearing house for ideas with interdisciplinary meetings held every two years. Publications of SEED include many booklets, one of which is a guide to design teaching.

Based on the German structured design approach, the Engineering Department at Cambridge University has completed a major revision of design teaching, has assisted in the translation of a number of German design publications, and has initiated several design research projects of which the one undertaken by Hales (1987) was most significant and will be discussed in Chapter 3.

In Switzerland, the prime aim has been the development of a theory of engineering design (Hubka, 1987) based, in turn, on a theory of the designed product, or *technical system*. This set of theories is intended to lead to a design science defined as a systematized body of knowledge about the goals, processes, procedures, techniques and objects of engineering design. In essence, this involves detailed classification and categorization of design processes and objects.

In Poland, design science and research has flourished for decades without much attention being paid by the West until recently (Gasparski, 1989). The principal roots of Polish research were in praxiology, a philosophical theory of action dating back to Kotarbinski (1965). Praxiology is also considered as a general methodology, i.e. the science of methods.

Design research was formally initiated at the Polish Academy of Sciences in 1969 when the Design Methodology Unit was established within the Praxiology Department, under the direction of Gasparski. Its objective was – as in Switzerland – to develop a design science but related to praxiology as proposed by Kotarbinski. This research resulted in establishing an active design community, in over 300 publications, including several books, and in the International Directory of Design Research (Archiszewski, 1990).

In praxiologic design science, three interrelated design elements are considered: the designers, the design object, and the design process itself. There are three sub-areas of praxiologic design science: (1) human aspects of engineering design, (2) engineering aspects of the design object, and (3) methodological aspects of the process. Particularly fruitful research was performed on the relationship between design education and design skills, and on the relationship between the designers' methodological knowledge and the relevancy of their designs. This design research led to the development of design courses and special training programs for designers. During the last 20 years, four international conferences on design methodology and praxiologic design theory have been held in Poland.

In Germany, developments in design methodology have taken place since the 1920s. Pahl and Beitz (1984), in their classic book *Engineering Design*, give a historical review which is both informative and inspiring. Today, methodical design is an integral part of German engineering curricula. This is supplemented by the publication of a series of design guidelines by VDI (Society of German Engineers), culminating in Guideline 2221 (VDI, 1987) as an aid to practicing engineers. This document is discussed in detail in Chapter 3.

Research in design methodology is being carried out at major German technical universities in Berlin, Munich, Karlsruhe, Aachen, Darmstadt and Braunschweig. Design methodology is being increasingly integrated into CAD, leading to a more methodical approach in this area. Design research in Germany is heavily application oriented and industry supported. Government support comes primarily through DFG (German Research Associations). The VDI works closely with industry and the universities.

Current design research in Germany falls into five categories: (1) basic research, (2) application to CAD systems, (3) computer implementation of methodical design, (4) expert systems in design and (5) understanding thought processes in designing. The work of Pahl (Darmstadt University) and of Beitz (Berlin University) is particularly important and is treated in detail in Chapter 3.

1.3.3 *United States of America*

A historical review of design research in the USA was included in Section 1.1. Nevertheless, a major step in promoting design research in this country occurred in 1984 when the NSF undertook a complete reorganization of its Directorate for Engineering. One effect of the new organizational structure was the establishment of the Division of Design, Manufacturing and Computer Integrated Engineering. Included in the division is a new program that focuses on design theory and methodology. A ten-person steering committee was established to consider the areas in which research in engineering design was needed in chemical, civil, electrical and industrial engineering, and in computer science. The committee published its report *Goals and Priorities for Research in Design Theory and Methodology* (Rabins, 1986).

This study, conducted under the auspices of the American Society of Mechanical Engineers, identified and defined a new engineering discipline: design theory and methodology. Five areas of study concerning this discipline were listed: (1) conceptual design and innovation, (2) quantitative and systematic design methods, (3) intelligent and knowledge-based systems, (4) information integration and management in engineering design, and (5) human aspects in design.

Since then a number of research initiatives have emerged. For example, the Engineering Design Research Center was established at Carnegie-Mellon University in Pittsburgh aimed at reducing the time between initial design and production by improving design tools and eliminating the need for intermediate steps to refine design. To achieve this goal, engineers must be able to model for reliability, safety and manufacturability.

Another research center, in Design Theory and Methodology, was established at Texas A & M University while a Design Division was named at Stanford University. The MIT Laboratory for Manufacturability and Productivity, directed by Nam P. Suh, and the Graduate School of Design at Harvard University are testimony to the renewed design thinking in the USA which has also been invigorated by the activities of the newly established Design Engineering Division of the American Society of Mechanical Engineers and the Design in Engineering Education Division of the American Society for Engineering Education. Finally, the launching of a new journal *Research in Engineering Design* (Springer Verlag) provides an important forum for the dissemination of research findings.

Regrettably, though, all these activities have by-passed the field of rock engineering and have not provided a much needed impetus in mining engineering. Only at the Pennsylvania State University has a concerted effort been made to promote design engineering in the mineral industries and to develop design principles and methodology specifically for rock engineering. The results of this effort and a firm conviction as to the importance of the subject have led to this volume.

In reviewing the current trends and the emergence of design engineering as a discipline on its own, it is appropriate to observe that many engineers in industry are still skeptical of the benefits provided by using a design process or are simply unaware of the many activities taking place in this area. In fact, it may even be too late or too unproductive to convince them, and major efforts would be better directed to improved design education for a new generation of design engineers.

CHAPTER 2

Creativity and innovation in engineering

To have the sense of creative activity is the greatest happiness and the proof of being alive.

Raymond Aron

An international conference on 'Creativity: the Essential Ingredient in Engineering Design', held in 1987 under the auspices of the American Society of Engineering Education (ASEE, 1987), was a landmark event in linking creativity, design and innovation. It demonstrated convincingly the crucial role played by creativity in engineering and pointed out the growing demand for creativity training. During the past year major national magazines, including *Newsweek, Business Week, Psychology Today, Omni, Success* and *Bostonia*, have written cover stories on the importance of creativity. Moreover, a visit to a major bookstore today will reveal over 20 books on the subject!

The most impressive evidence that creativity training has come of age can be seen in the latest industry survey by *Training* magazine. It reports that 25.6% of all organizations with 100 or more employees are now offering 'creativity' training. This figure is up from less than 4% in 1985 and represent a phenomenal 540% increase.

Creativity is important to design because the design process may be seen as a vehicle by which creativity is turned into innovation. *Creativity* is defined as using one's imagination to conceive novel concepts or original ideas, while *innovation* is implementing these concepts and ideas into new or improved products and systems. Thus, processing and developing creative ideas for practical use leads to innovation. The design process transforms creativity into innovation. People may be creative and knowledgeable but unless their original concepts are turned into innovation, these concepts will remain stillborn or somebody else will make use of them.

2.1 LINKING CREATIVITY, DESIGN AND INNOVATION

Much research has been devoted to creativity and innovation (Adams, 1988; Nadler and Hibino, 1990). Many of the earlier publications on creativity emphasize fairly narrow and mechanistic interpretations of mental effort or specific

15

techniques of creativity technology such as brainstorming. Weisberg (1986) presented a critical analysis of the many widely accepted notions and argued for the incremental nature of creativity emphasizing the need for hard work and acquiring the relevant technical skills.

Research has shown that all people possess – to a greater or lesser extent – some creative abilities. The author's experience with students is that the majority show a marked aptitude in this area, in some cases amounting almost to a hunger craving to be satisfied. Certainly some have a greater creative flair that others, however, in most cases, enlightened teaching can develop it further. It may be that few possess the potential to become brilliant designers, but with proper guidance the great majority can at least become good designers.

Creativity can be developed but it requires practice, persistence and patience. Most of all, it requires willingness to accept a challenge. It also calls for observing others, asking questions, and listening carefully to answers. It requires being prepared because one never knows when a good idea may come, so a pad and pen should always be kept handy!

One of the reasons why creativity is often not put into design practice is because some people and organizations tend to shoot down ideas. Here is a list of typical statements that are often aimed at new ideas:

1. It will never work.
2. We've never done it that way before.
3. We are doing fine without it.
4. We can't afford it.
5. We are not ready for it.
6. It is not our responsibility.
7. It was not invented here.

Some established companies and design managers have an unfortunate tendency to say 'no' to ideas that might be radically new or upsetting to their usual way of doing business. Then someone comes along, picks up the same idea and makes it work. So the first step in creativity training is: cultivate new ideas!

If the above list of the barriers to creative design were not enough, below is another list of a 'dirty dozen' of phrases collected by Smalley (1986):

1. We've tried that before.
2. We are too small (or too big) for that.
3. We have too many projects going on now.
4. What nitwit thought that one up?
5. Let's wait and see.
6. What's wrong with the way we do it now?
7. It will mean more work.
8. Sounds good, but it's too risky.
9. Interesting idea, but it won't work here.
10. I like it, but probably nobody else will.

11. Why something new now? Things are so comfortable.

12. You've got to be kidding!

It is generally accepted that each person has a hidden creative ability which is not being utilized. Why are we not using our creative talents to the fullest? One possible explanation is the 'split brain' theory. With reference to Figure 2.1, Williamson and Hudspeth (1978) explain: 'Psychologists have shown that the left hemisphere of the brain tends to concentrate verbal and logical reasoning, while the right hemisphere tends to concentrate visual and creative reasoning. Unfortunately, educational processes have become dominated by left-brain learning. It is quicker, easier, and more measurable than right-brain learning. It fits closer to learning objectives. Advanced degrees tend to select students most successful at left-brain operations. A transformation that should be facilitated in the education of any modern engineering student involves a shift from the singular dependence upon the left brain with its emphasis on the equation for the right answer to also include the right-brain with its ability of holistic analysis and synthesis so essential for creative design.'

The crucial role of creativity in engineering design is particularly manifested in three major design activities. One is *problem definition* from a 'fuzzy' array of facts and myths into a coherent statement of the question (Suh, 1990). It is one of the most important steps in design and is often done through an iterative process involving even the complete cycle of design as discussed in Chapter 3. Problem definition requires a lot of imagination, the essence of creativity. The second design activity where creativity plays a major role – and a decisive one at that – is,

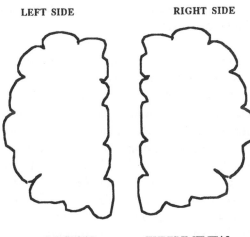

LEFT SIDE   RIGHT SIDE

LOGICAL   EXPERIMENTAL
DECISIVE   HOLISTIC, GLOBAL
ORDERED   IMAGINATIVE   Figure 2.1. The concept of the left and
DISCIPLINED   DAY-DREAMING   the right brain hemispheres used in
ANALYTICAL   CREATIVE   describing the process of thinking.

of course, the process of devising a proposed physical embodiment of the design solution. This is an *ideation* process and is highly subjective (Suh, 1990). The creative ideas and the synthesizing of a design solution into a physical embodiment depend strongly on the designers' creativity and knowledge base and their ability to integrate this knowledge. Therefore there can be an infinite number of possible creative solutions that can be synthesized to satisfy the design objectives: if we give the same functional requirements for the design of a product to ten designers, we may receive at least ten different design solutions for the product!

The third design activity where creativity is most relevant is the design evaluation stage of determining which is the best design and whether the proposed solution is appropriate or rational. This process of analyzing a design implies making correct design decisions as well as evaluating the details of specific design features. Again, there is a lot of scope for creative imagination in this design activity. Suh (1990) points out some questions which need to be answered: 'How do you make design decisions? Why is this design better than others? Is this decision rational?'

Chaplin (1989) discussed the urgent need for improved methods which could effectively develop and enhance the creative potential of young design engineers. His study indicated that an innovative product design is a creative undertaking requiring sensitivity and intellect. Creative design is stimulated by interest and curiosity. The designer must be fascinated by the design job and see it as an intellectual challenge. Creativity in the design process is needed not only at the conceptual stage, but right through to the detailing phase. Moreover, the designer needs to be aware of the historical development of the subject to see the current problem in perspective and as a source of creative ideas.

There are a number of techniques which may be used by designers to improve their creativity. Some of the most common ones are:

Brainstorming. Probably the most popular creativity technique, brainstorming involves a session where members of a group (or individuals working alone) are encouraged to create freely as many solutions to a problem as they can. Although participants are encouraged to suggest how their own ideas and those of others can be joined together to form a solution, no criticism or evaluation of their own or other ideas is allowed. The assumption under brainstorming is that the greater the number of ideas produced, the greater the probability of achieving an effective solution to a problem. Without criticism or evaluation, participants can concentrate more on determining and shaping solutions than defending their ideas. Furthermore, because brainstorming excludes factors which tend to induce emotion, it not only breeds self-confidence but stimulates creative thinking as well. Brainstorming works because it helps solve problems, encourages participation, promotes teamwork, and evokes a sense of purpose. Most of all, however, brainstorming makes creativity contagious. A single word can fire up the whole group or a single phrase can release a flood of associated thoughts and concepts. Examples of brainstorming sessions are given in Chapter 7.

Programmed Invention. One of the primary disadvantages of brainstorming is the possibility that someone will get credit for another person's ideas because an original thought may be lost to an improved idea. Programmed Invention (PI) attempts to correct this defect. The PI session is conducted very much like brainstorming in which ideas are wanted in quantity with 'wild' creative thinking encouraged. The PI session differs, however, in one important step. When a member of the group suggests an original thought all idea generation stops upon a signal from the group leader and every member focuses on the initial idea to improve, extend, and build on it. The originator of the idea records all pertinent constructive extensions of the idea on index cards. In this way the group truly works for the individual who gets full credit for the original thought.

Synectics. This technique of group idea generation or solving problems in a creative way was invented by William J. J. Gordon around 1960 (Adams, 1988). It utilizes metaphor to generate creative ideas. *Synectics* is a Greek word meaning the fitting together of seemingly diverse elements. A group of carefully selected individuals of varying personalities and areas of specialization is chosen with a leader playing a dominant role during the discussion. The problem or task is first explained to the group in detail. The leader then selects a method of attack such as presenting an analogous situation which may or may not have an obvious bearing on the problem. When an interesting idea of possible significance is suggested by someone in the group, the leader steers the discussion into an elaboration of and an analysis of the idea. In general, the technique of synectics is based on the fact that the mind is more productive when dealing with a new or foreign environment. Ideas are often drawn 'by association:' the analogous situation quickly takes one away from the exact problem at hand (with traditional approaches to a solution) and requires the designers to consider a related problem. This has a tendency to make the strange familiar or, in another situation when appropriate, the familiar strange.

The above three techniques are particularly effective for generating creative ideas. However, there are many others (see Chapter 7). Some organizations are even highly creative in stimulating creative ideas! *The Wall Street Journal* (December 9, 1988) featured a story about a Japanese company that paid its employees to network with people in professions unrelated to their own. Employees were reimbursed for the cost of taking these people to breakfast, lunch or dinner on condition that they learned just one new idea that would help them improve their own jobs! The company, Sumitomo Trust and Banking, recognized that its employees could learn from people in other professions.

In summary, we have clearly established that creativity is inseparably linked to engineering design which in turn transforms a creative idea into a practical innovation. One must therefore differentiate between creativity and innovation which entails carrying ideas through into something that works and makes a profit.

Well, what are the strategies of the most innovative companies? Firstly, they realize that company financed research and development are much more effective in generating industrial growth than federally financed R&D (Cohen, 1988). Secondly, their management style assures that R&D, manufacturing and marketing functions are closely knit and communicate well with one another (as is the practice in Japan). Thirdly, the education and training of technical employees receives a high priority; for example, Exxon and IBM have very comprehensive and high quality programs for technical education, career development and improving management skills. Moreover, intensive orientation programs are compulsory for new employees (in Japan, new employees go through a formal training period of 6 months to one year which includes, for technical personnel, such routine duties as in-plant operations or even retail sales). Fourthly, the emphasis on a quality process in product development extends not only to the service sector but also to research activities. In this last aspect, Exxon maintains that this shortens the time for commercialization by 30 to 50%. Fifthly and finally, successful industrial companies participate aggressively in university efforts to strengthen technology impact by sponsoring curriculum changes that will give students a good understanding of how the innovation process works (Steinbreder, 1988).

To help the designers in their creative and innovative endeavors, a number of techniques – discussed earlier – may be used to improve idea generation. But which techniques to use or even which approach may best suit a particular designer's personality and skills, is best decided upon once the very process of creative thinking is understood.

2.2 FUNDAMENTALS OF CREATIVE THINKING

Understanding the fundamentals of creativity is important to designers for many reasons, particularly because the latest research suggests that people can be taught to be more creative.

Isaksen (1987) pointed out three myths which often hamper people in their creative efforts. One myth is that creativity is a *mysterious* or mystical phenomenon because no one seems able to offer a universal definition or explanation. People who view creativity in this manner suggest that fruitful inquiry and discussion is not possible due to this ambiguity. Yet, there are over 200 scientific publications on creativity and there does appear to be a consensus on the nature of creativity among investigators most closely associated with work in the field. This includes four broad areas: the characteristics and attributes of the *creative person*; the criteria determining the *creative product*; the identification and description of the stages of the *creative process*; and the nature of the *creative environment*.

The second myth, according to Isaksen (1987), is that creativity is something *magical*. This implies that only a few people posses a special gift to be really creative. However, Simon (1985) – a recognized leading psychologist – rejected the idea that sparks of genius need to be present for creativity to exist: 'creativity appears to be accessible to everyone with a modicum of ability; there is an infrequency of extremely high level creativeness but if a person is able to think and solve problems then there does appear to be room for creativity.'

The final myth surrounding creativity is that in order to be creative a person must be *mad*. This myth asserts that creativity is based on the psychological processes of neurosis: that it is a function of a troubled mind. Accordingly, creativity is something to be avoided like any other form of sickness. Yet, although much popular literature seems to focus on creativity as madness, it is now widely believed that creativity is related to the natural development of human potential and that releasing creativity is healthy (Isaksen, 1987).

Does this mean we can teach creativity? Research on nurturing creativity shows (Gryskiewicz, 1980) that there are three main sources for becoming creative: (1) inherent abilities, (2) motivation, and (3) acquired skills. A combination of all three aspects can be particularly effective in developing creative behavior but even two sources are enough. This clearly shows that any willing engineering designer can indeed achieve a high level of creativity and break down rigid thinking that blocks new ideas.

An important point to remember, however, is that creativity must be cultivated because a great idea can be shot down in a matter of seconds; research shows (Isaksen, 1987) that for a new idea to succeed a praise/criticism ratio of as much as 4:1 is needed!

Accordingly, since creativity can be identified and assessed as well as nurtured, then it is important for anyone concerned with design innovation to develop their own creative problem solving skills and carefully select those with whom they work. Chapter 7 is devoted to this topic using the motto: *one can learn to be creative* (each year, over 20,000 corporate executives attend creativity workshops and there are some 100 consultants who teach creative thinking methods, including Harvard and Stanford, who offer creativity courses to business students).

Recent research has radically changed the way scientists think about creativity: the creative process can now be explained (Simon, 1985). By the 1970s many researchers were convinced that the key to creativity in the arts and sciences – the ability to discover new relationships and to look at subjects from new perspectives – was a property of the right side of the brain while the capacity for logical thought resided in the left side (see Figure 2.1) (Kerley, 1986). Those distinctions are now considered simplified and misleading. It was found instead (Smith, 1985) that creativity is 'a feat of mental gymnastics engaging the conscious and subconscious parts of the brain. It draws on everything from knowledge, logic, imagination, and intuition to the ability to see connections and distinctions between ideas and things.' Brain scanning, using a technique called tomography, shows that both sides of the brain flicker on and off when a person is engaged in creative thought. It has been said that 'creativity requires honing one or more of the intellectual processes to a high degree.' In essence, the seat of creativity is the sub-conscious mind which constantly analyzes and reorganizes ideas leading to creative thoughts or acts.

William E. Herrmann is one of the leading creativity consultants whose Whole Brain Corporation has a roster of 250 client firms (including General Electric, Procter & Gamble, Polaroid and Shell) and annual revenues of $1.8 million. He proposed (Smith, 1985) that 'creativity is not a single talent but a combination of different types of thinking – analytical, verbal, intuitive, and emotional – each controlled by a separate region of the brain.' Only about 5% of individuals show equal development in the various types of thinking (most of them are chief executive officers). So, he devised a test that allows him to determine which type of thought is dominant in a person. He then assembles groups in which various members represent all the critical talents for creativity – in effect, forming a *whole brain.*

The latest research (McAleer, 1989) also identified some common traits that characterize a creative thinker:

1. Creative types are generally independent, persistent, and highly motivated. They are also great skeptics, risk takers, can be hard to get along with, but they do have a strong sense of humor.

2. Creative people usually do not have dull, predictable childhoods. Instead, their childhood was marked by exposure to unusual events and freedom in decision making.

3. The most gifted creators are not loners as popularly believed; they are constantly exchanging ideas with colleagues at the cutting edge of their fields. Studies have shown (Smith, 1985) that scientists and engineers who talk most with their peers publish more papers and produce more innovative work than their more aloof colleagues.

4. Some college education may foster creativity but in some fields it perpetuates entrenched thinking. Modern education, which stresses logic, seems to squelch creativity. Tests show that a child's creativity plummets 90% between ages 5 and 7. By the age of 40 most adults are about 2% as creative as they were at 5. The intelligence threshold for creativity is an IQ of about 130. Above that, IQ does not make much difference.

5. Creative people are highly objective. They not only scrutinize and judge their ideas or projects, but they also seek criticism.

6. Nearly all people who are renowned for creativity spent years mastering their field: in one study, it took most great musicians and painters 10 years of hard work before they produced a masterpiece. In science, Thomas Edison had over 1,000 inventions to his credit before succeeding with a few great ones, and resented being called a genius which he defined as *99% perspiration and 1% inspiration.*

Chaplin (1989) concluded that creative thinking, that is, the formation of original ideas or original combinations of ideas is essential to effective and competitive design. The nature or style of creative thinking varies throughout the design process; a different kind of creative thinking being required at the early conceptual stages of design from that needed at the detailing stage. A thorough and detailed analysis of a problem is an essential precursor. Creative thinking benefits from the right environment, and needs the right attitude of mind. It is helped by self-confidence, persistence and determination, an adventurous spirit, and a receptive, open mind. It is hindered by criticism and judgment, inability to break out of a set pattern, and by timidity.

The most important conclusion after considering the fundamentals of creativity is that creative thinking can be learned and improved by practice and experience thus greatly aiding designers in their quest for innovative solutions. Understanding the creative process is important because creative efficiency in people can be markedly increased if they understand the psychological processes by which they operate. Thus, an insight into how designers design is particularly beneficial.

2.3 TURNING CREATIVITY INTO INNOVATION: HOW DESIGNERS DESIGN

The very process by which humans create their designs fascinated, puzzled and frustrated scores of writers, researchers and even designers themselves. This search for the 'golden key' to unlocking the mysteries of successful designers continues to this day because some people design without any methodology while others are highly organized; most of all, even accomplished designers have problems in describing their own design actions. Accordingly, this subject has received considerable attention in various disciplines: from fine arts through architecture to engineering. In fact, a whole field of 'protocol analysis' has emerged with researchers subjecting designers to a close scrutiny while engaged in the process of design, not unlike the 'time-and-motion studies' of the past. Certain views have by now been formulated as to how designers design – in engineering and other fields.

2.3.1 *Engineering designers*

Finger and Dixon (1989) presented a thorough review of descriptive and prescriptive models of engineering design. They divided the descriptive aspect into two categories: one that gathers data on how designers design and the other that builds models of the cognitive process. A cognitive model is a model that describes, simulates or emulates the mental processes used by a designer while creating a design.

A background to the nature of protocol studies of designers was given by

Finger and Dixon (1989) as follows: 'Much of the design process is a mental process; the sketches and drawings that form the visible record of designs do not disclose the underlying processes by which they were created. In a design protocol, the actions of a person performing a design task are recorded as the design evolves. Usually, the designer is encouraged to think aloud and is questioned when information seems to be missing or incomplete. Because there is no single design process or design strategy, most protocol studies are set up to study a few well-defined questions. For example, the strategies used by expert designers performing a familiar design task might be compared with their performing an unfamiliar task.'

It was found that while there is a consensus that designers exhibit reusage of design strategies during all phases of design, this is an assumption that has never been tested. In addition, few formal protocol studies have been done on design teams. Finger and Dixon (1989) pointed out that one of the major criticisms of design protocols is that a designer's word cannot reveal those processes that are inherently non verbal, for example, geometric reasoning. Moreover, the requirement to verbalize may interfere with the design process itself.

Adelson (1989) discussed her approach to cognitive research, defined as uncovering how designers design, and to cognitive modeling defined as explaining and predicting how designers design. Her paper is a good introduction to protocol studies.

Using protocol analysis (Ericsson and Simon, 1984), Ullman and Dietterich (1986) have performed a study of engineering designers looking at novice and expert designers designing mass-produced and one-off products. They concluded that designers pursue a single design concept, and that they will patch and repair their original idea rather than generate new alternatives. This single-concept design strategy, which has also been observed in software designers, does not conform to the traditional view of what the design process ought to be as discussed in the previous section.

Stauffer and Ullman (1988) compared six investigations in which the mechanical design process was evaluated by studying human designers. They concluded that while many researchers make a case for an algorithmic view of design (a specific sequence of steps to solve design problems), the actual design performance is not that well organized. In many cases there was no strategic plan in design and the designers observed in the studies did not follow any set procedures; any procedures that may have been followed were general, and the designer's attention usually shifted to critical parts of the problem or 'became opportunistic.' In essence, the designers followed 'rules-of-thumb' to solve problems, which were dependent on the situation at hand. It was further concluded that a systematic design approach could only improve design, in the absence of which an unstructured heuristic approach to design is actually performed by designers. Table 2.1 summarizes all the conclusions by Stauffer and Ullman (1988). Note that these conclusions are arranged in four topics for easier understanding: systematic

Table 2.1. How engineering designers design: Conclusions from a study by Stauffer and Ullman (1988).

1. Engineers commonly employ means-end analysis (use of heuristics).
2. Engineers commonly make a qualitative plan for designing without specifying the details.
3. The initial focus of the conceptual design is on critical areas.
4. The design process is multidirectional, and there is no clear distinction between conceptual, layout, and detail design phases.
5. Design is sometimes opportunistic rather than systematic.
6. Designers do not always conduct balanced development, but sometimes pursue a problem in a depth-first manner.
7. Designers develop the functional aspects of the design in stages throughout the problem-solving effort.
8. Designers make decisions based on qualitative, subjective reasoning.
9. Designers evaluate the solutions to a problem by evaluating their respective sub-problems
10. Goals are formulated during the problem-solving process.
11. The design goals are initially qualitative, but become quantitative as the design progresses.
12. Designers use functional considerations that remain qualitative.
13. Problem solutions occur in parallel when designing in teams; decisions are made more quickly.
14. Problem solutions occur in series for single designers working alone or when problem is difficult.
15. Engineers commonly accept a solution that is satisfactory even if it does not represent an optimal result.
16. Design is both a technical and a behavioral process.
17. Design is influenced not only by the personal abilities of the designer and the client, but also by how they interact.
18. The type of organization that employs the design engineer and how it operates has an effect on design.
19. Initial premises for concept generation are often false.
20. Individual designers have favorite solutions.
21. Simulations are made to help evaluate a problem.
22. Knowledge from diverse sources and different individuals is 'integrated' and 'contributes to the collective knowledge base used in conceptual design'.
23. Analogies are used as basis for concepts.
24. Models are used to set design parameters and make configuration decisions.
25. Designers use notes and drawings for understanding and analyzing the problem.
26. Designers use their knowledge to influence how ideas are generated and evaluated.
27. Designers use their knowledge to influence their problem-solving methods: the design process *is* dependent on the domain knowledge.

(algorithmic) versus heuristic nature of design (items 1-12); parallel versus serial development of solutions (items 13-15); the technical versus behavioral nature of design (items 16-20); and the dependence versus independence of the design process on domain knowledge (items 21-27).

Hales (1987) performed a detailed case study on designers in action based on 2.8 years of observation of a single design project encompassing the work of 37 individuals in Britain, ranging from the design engineers to staff responsible for marketing and management. He emphasized the importance of distinguishing between the design process and the context within which the process takes place.

Bucciarelli (1988) reported two studies of designers within engineering firms based on participant-observation technique. He concluded that design is a social process in which different designers think about the work on the design in quite different ways.

Dixon (1991), discussing descriptive design models, concluded that very little is known about the cognitive processes by which engineering design is done. It is known that designers with more experience can create better designs because their knowledge is broader, more general and more abstract. It is also known that most individual designers reuse familiar solutions and do not explore alternatives well or thoroughly; instead they try to 'fix' their original ideas. Most of all, the processes of designers working in groups have only been studied to a very limited extent.

In mining, a study of how engineers design by Sanders and Peay (1988) concluded: *The engineers relied overwhelmingly on design solutions that they used before. This prevented them from considering novel approaches to the design. Further, there was a reluctance to modify their initial designs, when new information became available.*

To summarize, studies of how designers design indicate (Whitney, 1990) that 'design practice has evolved right out from underneath the feet of researchers driven by the needs of advanced industries and technologies.' In most firms today, design is not limited to those who are educated as designers or who spend most of their time designing products or processes. Many more engineers and scientists participate in design than those whose job assignments are design. NRC (1991) presented these observations about the attributes of excellent designers:

1. Effective engineering designers seem to have great associative power that lets them recognize and draw upon parallels in other fields for ideas.

2. Good designers presented with a problem always seem to respond with a flood of ideas rather than a single solution; they are highly creative.

3. Good design engineers often have strong inner-directed personalities; being sure of their own worth and contribution, they are able to accept that sometimes a poor solution will creep in which will not diminish their high reputation earned from so many great solutions. Thus they are more adventurous when seeking solutions.

4. Top engineering designers are very productive: a large fraction of the design in a firm may often be done by a small fraction of the most effective designers.

Cross (1989) reported that when designers were asked to discuss their abilities and explain how they work, a number of common themes emerged. One theme was the importance of creativity and intuition in design, even in engineering design. Another theme was that 'the solution' is not always a straightforward answer to 'the problem' but may come about from 'round-about' considerations. A third common theme was the need to use sketches, drawings or models to explore the solution.

Jansson (1989) discussed an important finding from a series of experiments which were conducted to test the hypothesis that *design fixation*, defined as a blind adherence to a set of ideas or concepts limiting the output of conceptual design, is a measurable barrier in the conceptual design process. The results of the experiment, which included professional engineers, clearly demonstrated the existence of design fixation. In other words, when one group of designers were given tasks with a sample design while the other group was given the same task but without an accompanying example, the range of design ideas was severely restricted in the group receiving the sample design. Most of the designers produced some variations on the theme of the design sample. In fact, the 'fixation group' of designers even conformed to the negative features of the design example. Thus, design fixation has potentially detrimental effects on designers: it may limit creativity in the design process. However, it is also possible that, in practice, design fixation might have some beneficial effects, such as selection of a good design feature from a past design solution. After all, the nature of conceptual design is such that 'prior information' is an essential element of the process. Designers can see new configurations based on what they know, not on things which they do not know. Accordingly, Jansson (1989) recommends that design fixation, as a barrier to progress in conceptual design, if it is to be a useful concept, should be viewed only as that which prevents the consideration of all the relevant knowledge and experience which should be brought to bear on any given problem.

Finally, Petroski (1991) summarized well the reality of how engineering design is created: 'A first design may be a triumph of elegance and individuality, but it is the drudgery and collective effort of the many designers, detailers, analysts, contractors, inspectors, and a host of additional experts and specialists that make engineering designs work and work safely.'

2.3.2 *Non-engineering designers*

Analyzing the performance of outstanding designers in non-engineering fields can be both informative and meaningful for engineering designers. Particularly the field of architecture abounds not only with information on how architects design (Mitchell, 1990) but also with comparisons between architectural and engineering designers (Bucciarelli et al., 1987). The latter reference showed that despite the difference between engineers and architects – in the objects, in styles of working, in rewards and values – there are structural similarities in the way designing proceeds. Although architecture is concerned with both function and form, while engineering is commonly viewed as an entirely functional activity that draws upon science and economics to produce solutions to technical problems, these formulations are inadequate because there *are* aesthetic considerations in the design practice of engineers. After all, bridges or dams may be seen not only as functional structures but also as works of elegance and aesthetic appeal in their own right. However, there is one essential difference: architectural criticism is a recognized and flourishing occupation; there is no such profession in engineering. This is one reason why creative design and elegance are stressed in architecture.

Great designers in architecture and landscape design have been studied but their secrets are not easily deciphered. Arthur Bye, a man who is considered by his professional colleagues one of the best landscape architects in America, has agreed to a study of his design process and his philosophy. However, he is an intuitive designer who often develops plans in a few seconds and works without detailed drawings. His biographer summarized Bye's design process: *A lot of what he does is done right in the field. He will have a bulldozer doing his composing, moving things around until he gets what he wants.*

Glegg (1969) reported that in a selection of 15 creative artists in various fields from music to painting, their inspiration came suddenly and unexpectedly and never when they were working at it! Best creative ideas occurred when: half asleep in bed, out walking or riding, in church or sitting in front of the fire. To this, the author would like to add his own sources for design inspiration: while at a symphony concert, or in the bathtub! Apparently, concentration and then relaxation is the common pattern behind most creative thinking.

In summary, investigations of what it is that designers actually do when they are designing is an important topic. An objective understanding of how designers design might in turn lead to the development of improved design procedures. Cross (1984), reviewing four in-depth studies of the nature of design activity, concluded that in the light of observations of how designers design, the systematic analysis-synthesis procedure seems ill-matched to the conventional design process particularly as practiced in architecture. But studies of designers may also lead to misleading conclusions: designers might post-rationalize on their activities, have faulty memories, and certainly can have difficulty in describing non-verbal design processes in words. Protocol studies have their own weaknesses: they are mostly performed by industrial psychologists without design experience and seldom involve design teams working under the conditions of pressure and competition found in the industrial world. What is needed are 'participatory' observations of designers in action where an experienced designer is a member of the design team and is willing and able to document the overall design experience and to explain the particular design actions. This, of course, places severe demands on the 'participatory' designer's time and it is not surprising that few case histories of this nature have been reported. Chapter 6 presents a number of cases from the author's own experience.

It should be emphasized, as Cross (1984) points out, that although conventional design practices in architecture as well as in engineering are primarily of an unstructured nature, the systematic design approach was conceived specifically to change this *ad hoc* design process which is inadequate for the complexity of the tasks facing modern designers.

Most of all, as *old habits die hard* and practicing designers may not wish to change their ways, a systematic design approach is a major opportunity for the new generation of designers. This systematic approach (described in the next chapter) clearly demonstrates that the design process is a *decision-making* process. Design is not just a reactive *problem-solving* activity but involves a series of considered decisions at various stages of a creative undertaking.

CHAPTER 3

Engineering design process

Scientists discover what is, engineers create what has never been.

T. von Karman (1911)

Not many centuries ago, the word 'engineer' was exclusively a military term. Civil, that is civilian, engineering had been invented by the French in the mid-18th century. The origin of the term *engineer* does not lie in the English term 'engine', as is popularly believed. It is derived from the French word *genie* meaning 'ingenious.' Hence ingenuity, creativity, and innovation should be the aspiration of all engineers.

The engineer's *genie* manifests itself in one word: *design*. It is design which makes engineers out of applied scientists and it is design that represents the culmination of all engineering training.

3.1 SCOPE AND NATURE OF ENGINEERING DESIGN

The importance of engineering design is underscored by the professional training requirements prescribed by the Accreditation Board for Engineering and Technology which offers this definition (ABET, 1987):

'Engineering design is the process of devising a system, component, or process to meet desired needs. It is a decision-making process (often iterative), in which the basic sciences, mathematics, and engineering sciences are applied to convert resources optimally to meet a stated objective. Among the fundamental elements of the design process are the establishment of objectives and criteria, synthesis, analysis, construction, testing and evaluation. Central to the process are the essential and complementary roles of synthesis and analysis. In addition, sociological, economic, aesthetic, legal and ethical considerations need to be included in the design process.'

It is clear from the above definition that the solution to any real engineering problem is never merely technological.

Nadler (1986) proposed a Systems Methodology and Design approach (SMD) which concentrates on defining the purpose of the system and on determining its level of complexity, thus significantly increasing the chance that engineers will be working on the right problem. *Define the problem and solutions will emerge* is a conventional saying based on the already existing context of data. The SMD approach emphasizes looking at the problem beyond its existing framework; it searches for what the objectives or problem ought to be. It also leads to improved productivity and more effective management. He pointed out that manufacturing output is much easier to measure than the output from the design office. A designer's productivity is measured qualitatively as well as quantitatively, meaning that the value of design work is related not only to the number of designs produced but also to their effectiveness.

Some problems detrimental to productivity and performance are the differing goals of managers and design engineers, often compounded by ineffective communication between the two groups. Other problems are the lack of a sufficiently broad and long-term perspective for a design project in the face of a desire for quick and visible results.

The SMD approach recognizes these characteristics of the nature of design: (1) *Design is hierarchical*, which is the key to understanding the complexity of levels and subsystems and how they are related; (2) *Design is functional* – it must lead to creation of a product which will perform a useful function in a satisfactory manner; (3) *Design means decision-making*, that is, selecting one course of action from several alternatives after careful evaluation; (4) *Design is iterative* which is necessary to co-ordinate, change, and improve upon previous decisions, designs and even goals and functions; (5) *Design seeks an optimal value* – rather than attempting to perfect each individual part within a system, the goal of design is to create an optimal whole system.

The SMD treats every problem as if it were new. Solutions to similar problems are not automatically adopted although this does not obviate the need to know about existing solutions. In essence, an SMD perspective visualizes the design engineer of the future as a combination of scientist, chemist, environmentalist, economist, sociologist, speaker, writer, and knowledgeable humanist. Most of all, the design engineer must learn to communicate more effectively in several modes.

The Systems Methodology and Design approach follows this process: determine the needed purpose(s) and the hierarchy of objectives; generate alternative ideas; develop an ideal solution; develop and detail practical solutions; carry out the proposed design and follow up.

Clearly, the essence of engineering is embodied in the design process by which the products are established. Accordingly, engineers should be recognized as much by how they go about designing as by what they design.

3.2 GENERAL DESIGN THEORIES

Engineering design involves the concepts of design theory and methodology. *Design theory* refers to systematic statements of principles and experimentally verified relationships that explain the design process and provide the fundamental understanding necessary to create a useful methodology for design. *Design methodology* is the collection of procedures, tools, and techniques that the designer can use in applying design theory to design.

Note that to claim a *design theory* one must have *systematic statements of principles*. They facilitate analysis and decision-making, and help the creative process of the design activity. Without them, design would be a mysterious creative process but with them, it is a rational and systematic activity.

The basic premise of a sound design philosophy should be that design is an inseparable part of the overall technological system of manufacturing or construction and provides the primary database for other activities that go on in that system. It is thus essential that the result of design must be realizable in the real world, otherwise the design would be but a dream. This realizing activity is manufacturing or construction.

Many researchers tried to generalize the designer's design activity and this proved useful for improving the design skills of professional engineers and for educating students. The outcome of the researchers' efforts was mostly composition of the elements of design activity into an innovative design process. This did not lead to design theories but to design methodologies which is discussed in the next section.

Design theories are like linguistic theories which explain the nature of the language but are not directed at changing it. Design theories are useful because they explain the mechanism of design and provide guiding principles for improved design knowledge collection, representation, and utilization. A significant contribution was made in the USA by Asimow (1962). He was mainly interested in the methodology of the design process and proposed an elementary representation depicted in Figure 3.1. Rodenacker (1970) of Germany made another important contribution to the theoretical aspects of design. His fundamental idea was that a system or a machine to be designed should be represented from the functional point of view. For example, a machine design necessitates three types of input: material, energy and signal (control), and exerts an output in the same terms, as shown in Figure 3.2. Thus the function of the machine is represented by the transforming properties of these three aspects.

Rallis (1963) of South Africa proposed a series of design steps in an inspiring paper on the importance of design as the goal of engineering activities. His work formed the foundation for the author's interest and subsequent research into the design process (Bieniawski, 1984).

Kotarbinski (1965) of Poland developed a general theory of systems related to a general style of problem solving leading to the methodology of design as a distinctive feature of the practical sciences. Design research was formally initiated at the Polish Academy of Sciences in 1969 when the Design Methodology Unit was established within the Praxiology Department, under the direction of

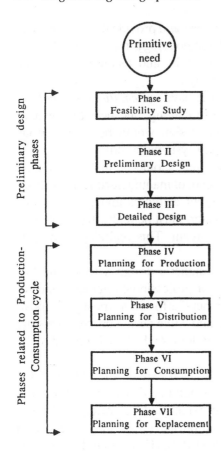

Figure 3.1. Design morphology proposed by Asimow (1962).

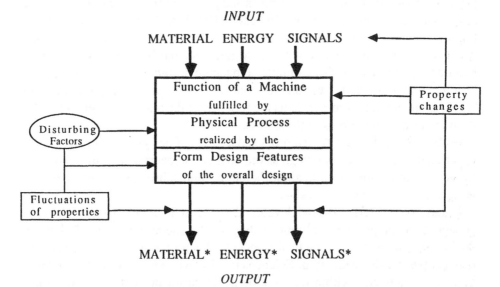

Figure 3.2. Structure of design functions visualized by Rodenacker (1970).

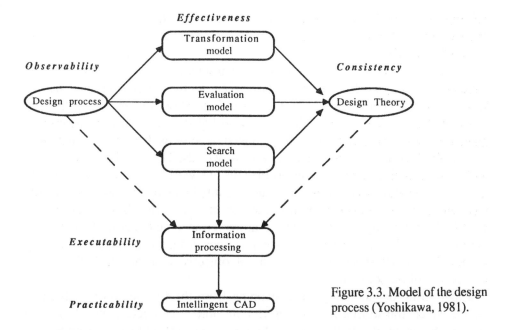

Figure 3.3. Model of the design process (Yoshikawa, 1981).

Gasparski (1989). Its objective was to develop a design science related to praxiology as proposed by Kotarbinski.

Building on the above and other researchers, three major contributions were made in the area of design theory: those by Yoshikawa (1981), by Hubka (1987), and by Suh (1990). Each approach was developed independently, and – ironically – with little regard to the other two. Yoshikawa believes that Hubka's and Suh's theories have the shortcomings of not being computer-implementable or computable. In his opinion, a design theory must describe the design process logically so that one can trace design actions using computers. Hubka does not even refer to either Yoshikawa or Suh. In turn, Suh pointed out the inadequacies of Yoshikawa's and Hubka's theories. He considered Yoshikawa's approach an algorithmic technique which applied design rules or classification methods to a specific situation. Hubka's theory is ignored by Suh although reference is made to Hubka and Eder (1988) as merely illustrating the use of design catalogs as a collection of known information for a specific problem.

However, all three originators of design theories agree on one point: the importance of understanding the design process. It is thus highly informative to examine in detail each of the three design theories.

3.2.1 *Yoshikawa's Design Theory*

The General Design Theory of Yoshikawa (1981) is based on a topological model of human intelligence (see Figure 3.3) and has three aims: (1) clarifying in a scientific way the human ability to design; (2) producing practical knowledge about design methodology; and (3) framing design knowledge in a form suitable for implementation on a computer.

As shown in Figure 3.4, from a philosophical standpoint, Yoshikawa visualizes three domains or 'worlds': the real world where all the concrete entities that we know exist, the conceptual world where we think about the entities of the real world, and the logical world which is the world of symbols, logic, mathematics, philosophy, etc. These three worlds are linked by various disciplines which exist as a result of 'mappings' between the various worlds. Thus, design theories deal with a mapping from a conceptual world to the real world via the logical world. In this case, designing is an activity to create an entity in the real world, from the first idea about the design object born in the conceptual world, through the logical world where drawings are made. Note that written specifications normally exist in the logical world. In essence, Yoshikawa defines design as a 'mapping' activity from a functional space (specifications of objectives) to an attributive space (properties of the solution).

Engineering or technology is a mapping from the logical world to the real world which transforms logical descriptions into a concrete entity. Here, the logical descriptions are the design drawings and an entity corresponds to a product of manufacturing.

In this visualization, design theory cannot be discussed without its engineering aspect, namely that designing includes a mapping from the logical world to the real world and not just only the mapping from the conceptual world to the logical world. The necessary condition for designing is an ability to form a concept about

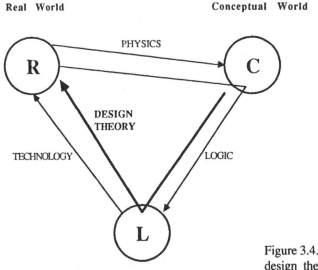

Figure 3.4. The concept of Yoshikawa's design theory (Tomiyama and Yoshikawa, 1987).

non-existent things as a result of using knowledge about existing things. With this background in mind, Yoshikawa (1981) provided these definitions:

Definition 1: The entity set is a set which includes all entities in it as elements.

Definition 2: Attributes of an entity are the properties: physical, chemical, mechanical, geometrical or other property that can be observed by scientific means.

Definition 3: When an entity is exposed to a circumstance, a behavior of the entity is observed which corresponds to that circumstance. This behavior is called a visible function. Different behaviors are observed for different circumstances.

Definition 4: A concept of entity is a concept that one has formed according to the actual experience of an entity.

Definition 5: An abstract concept is derived by the classification of concepts of entity according to the meaning or the value of entities.

By using these definitions, the General Design Theory establishes three axioms:

Axiom 1 (Axiom of recognition): Any entity can be recognized or described by its attributes.

Axiom 2 (Axiom of correspondence): The entity set in the real world and the set of entity concepts (ideas) have one-to-one correspondence.

Axiom 3 (Axiom of operation): The set of abstract concepts is a topology of the sets of entity concept. It is possible to operate abstract concepts logically as if they were ordinary mathematical sets.

From these axioms one can deduce theorems which describe the designer's designing process:

Theorem 1: The design requirement is the intersection of abstract concepts in functional space.

Theorem 2: The design requirement is a filter.

Theorem 3: The design solution is the intersection of abstract concepts in attributive space.

Theorem 4: If design is possible, the identity mapping from the attributive space to the functional space is continuous.

It is obvious from the above that the General Design Theory of Yoshikawa is not easy to follow or to apply. Apparently, it is particularly promising for CAD/CAM applications and the interested reader is advised to consult the original paper by Yoshikawa (1981) and a recent one by Tomiyama and Yoshikawa (1987). A 'translation' of some of the concepts may also be helpful:

- entity: real object, existing or to be designed
- attribute: property of an object or an assigned value
- function: behavior of an object under certain conditions
- concept: idea or perception of entities, attributes or functions
- functional space: specifications for a product to be designed
- attributive space: characteristics of the solution.

There are four ways of relating functional space to attributive space (this 'mapping' process is defined as design) for purposes of automating design (CAD/CAM):

1. One-to-one correspondence (the design process is a choice of existing solutions).

2. Calculation model (all attributes have numerical values and it is possible to calculate the solution from specifications mathematically).

3. Production model (a number of generally valid production rules can be applied to transform specifications into a solution).

4. Paradigm model (the designer finds one solution which seems to satisfy some of the specifications and then modifies it for better and better performance until all specifications are met.

3.2.2 *Hubka's Design Theory*

The 'theory of technical systems' of Hubka (1987) was originally published in German with an English translation by Hubka and Eder (1988). The primary aim of this theory is to classify and categorize knowledge about 'technical systems'

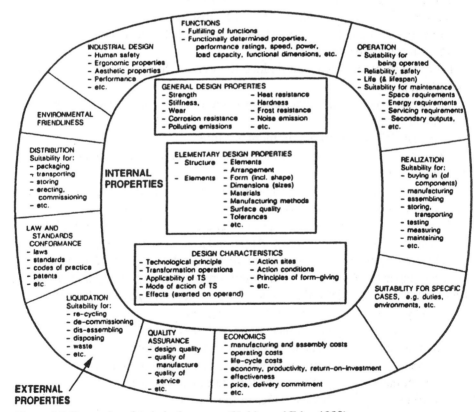

Figure 3.5. Properties of technical systems (Hubka and Eder, 1988).

(design objects or processes) into an ordered set of statements about their nature, regularities of conformation, origin development, or various empirical observations. At the same time, a suitable terminology is created such that their meanings do not need further explanation.

Hubka considers designing a particularly fruitful domain for the theory of technical systems. He defines engineering design as a process performed by humans aided by technical means through which information in the form of requirements is converted into information in the form of descriptions of technical systems, such that this technical system meets the requirements of mankind.

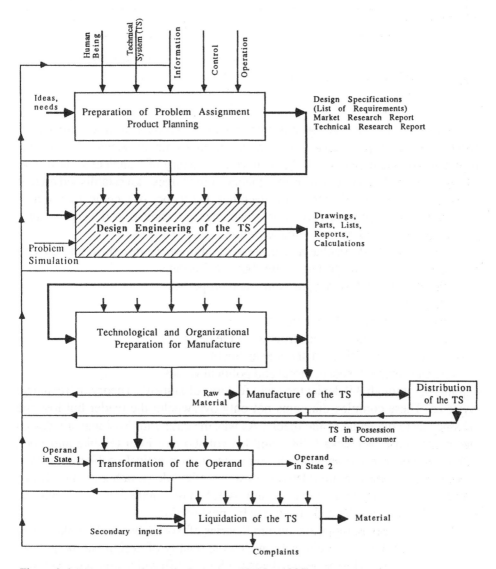

Figure 3.6. Life stages of technical systems (Hubka, 1987).

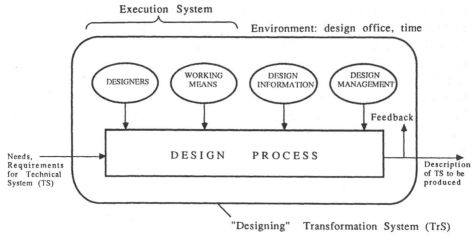

Figure 3.7. Model of Hubka's design process (Hubka and Eder, 1988).

In other words, 'the designer designs a device, product, artifact, process or procedure in a perceived context (organization, environment, society).' Engineering design is visualized as a bridge between the theory of technical systems and design science. The same way as the 'theory of machines,' an established area of knowledge, serves the designers of machines, the theory of technical systems – which are objects, products, processes, machines or technical works and means – can help design engineers in a broader sense.

Hubka (1987) believes that a discussion of engineering design should cover a number of factors as suggested by the above definition:
- the object to be designed, its nature and properties
- designers
- working means: tools and aids
- the activity: sequence and structure of the design process
- the context: organization and management
- the environment: social, moral and political values.

Figures 3.5 through 3.9 depict the main features of Hubka's Theory of Technical Systems. These complex diagrams not only overwhelm the reader but lose the message as to the value of the theory. As was the case with Yoshikawa's design theory, the work of Hubka (1987) and his collaborator Eder (Hubka and Eder, 1988) are not readily applicable to the solution of practical problems. In fact, some of their work is more related to design methodology, rather than to design theory, but is not as comprehensive and useful as the methodology of Pahl and Beitz (1984) which is treated in detail in the next section.

The main shortcoming of the design theories of Yoshikawa and Hubka is that they lack clear design principles by which the quality of design can be judged. This is not the case with the design theory of Suh (1990) which must be given sole distinction as the most workable design theory.

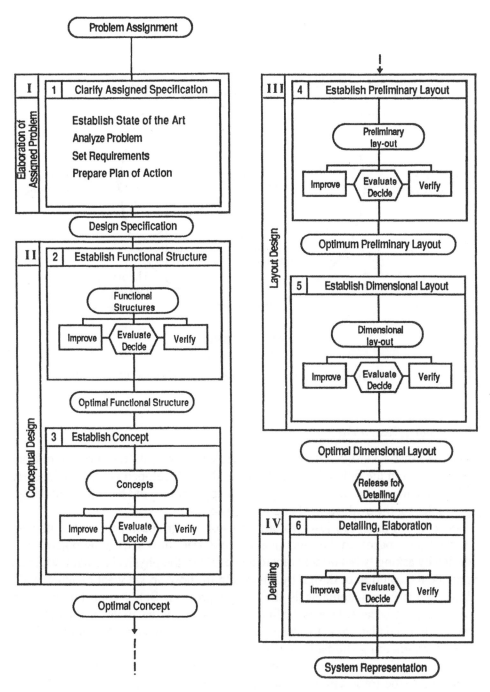

Figure 3.8. Model of methodical procedure during design (Hubka, 1987).

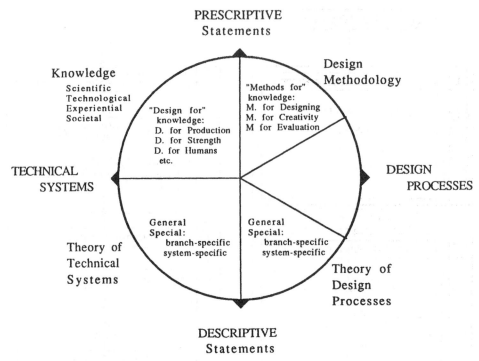

Figure 3.9. 'Two dimensions' of design theory, according to Hubka and Eder (1988).

3.2.3 *Suh's Design Theory*

The Axiomatic Design Theory of Suh (1990) represents the most significant contribution in the field of design theory to date. Developed at MIT in the early 1980s, the theory has matured for practical applications and has provided design principles for the evaluation of engineering design.

Suh (1990) proposed just two principles of design, each pertinent to its own domain (i.e. space). In the functional domain, one must satisfy the objective of design by asking 'what do we want to achieve'? In the physical domain, we must provide the solution of design by answering to 'how do we want to achieve it'? Linking these two domains is the design process. Suh's contribution is an important one because he was the first to suggest analytical tools for evaluation of the synthesized ideas so as to enable the selection of only good ideas and to offer a basis for comparing alternative designs.

The two principles which govern Suh's design process are: the Independence Axiom and the Information Axiom. The Independence Axiom is related to the functional domain and states: 'Maintain independence of functional requirements (FR).' The Information Axiom is related to the physical domain and states: 'The best design contains the minimum of design parameters (DP) satisfying the corresponding FRs.' Ideally the number of functional requirements should equal the number of components of the design solution. 'Minimum information', in essence, means least complexity. The overall idea is that the best design would have a number of independent functional requirements which will be satisfied by the simplest solution, i.e. one featuring the fewest design components. A 'design parameter' (DP) represents a design solution.

Although there are only two axioms in the Suh design theory, he also provided a number of corollaries as well as theorems. Apparently, when this work started in 1977, the Suh group at MIT evolved 12 hypothetical axioms which were soon reduced to six axioms and six corollaries. However, further work led to a realization that the six hypothetical axioms could be further reduced to just two axioms. Although they planned to add a few more, to this date they have not come up with any new axioms.

According to Suh, 'design may be formally defined as the creation of synthetized solutions in the form of products, processes or systems that satisfy perceived needs through the mapping between the FRs in the functional domain and the DPs of the physical domain, through the proper selection of DPs that satisfy FRs.' This mapping process is not unique; therefore, more than one design may ensue from the generation of the DPs that satisfy the FRs. In other words, the actual outcome depends on a designer's individual creative process. Accordingly, there can be an infinite number of plausible design solutions and mapping techniques. The design axioms provide the principles that the mapping technique must satisfy to produce a good design, and offer a basis for comparing and selecting designs (Suh, 1990).

From the two axioms, corollaries may be derived which will be useful in making specific design decisions since they can be applied to actual situations more readily than the original axioms. Suh (1990) states that these corollaries may even be called *design rules* and are all derived from the two basic axioms.

Corollary 1: Decoupling of coupled design. Decouple or separate parts or aspects of a solution if functional requirements become interdependent in the proposed design.

Corollary 2: Minimization of Functional Requirements. Minimize the number of FRs and constraints.

Corollary 3: Integration of physical parts. Integrate design features in a single physical part if functional requirements can be independently satisfied in the proposed solution.

Corollary 4: Use of standardization. Use standardized or interchangeable parts if the use of these parts is consistent with the functional requirements and constraints.

Corollary 5: Use of symmetry. Use symmetrical shapes and/or arrangement as if they are consistent with the functional requirements and constraints.

Corollary 6: Largest tolerance. Specify the largest allowable tolerance in stating functional requirements.

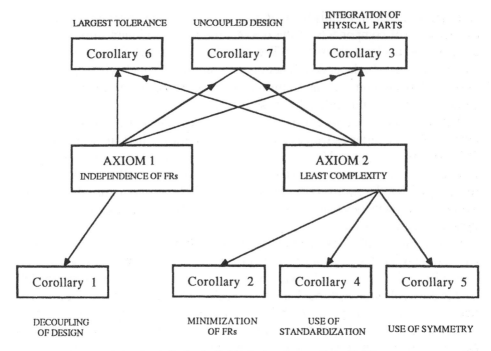

Figure 3.10. The origin of corollaries in Suh's design theory (Suh, 1990).

Corollary 7: Uncoupled design with less information. Seek an uncoupled design that requires less information than coupled designs in satisfying a set of functional requirements.

Figure 3.10 shows the relationship between these corollaries and the axioms. It shows that Corollary 1 is a direct consequence of Axiom 1, while Corollaries 3, 6, and 7 are derived from Axioms 1 and 2. Corollaries 2, 4, and 5 are derived from Axiom 2.

Other implications of these design axioms are that they lead to a number of theorems (propositions) listed by Suh (1990):

Theorem 1: Coupling due to insufficient number of DPs. When the number of DPs is less than the number of FRs, either a coupled design results or the functional requirements cannot be satisfied.

Theorem 2: Decoupling of coupled design. When a design is coupled due to the greater number of Functional Requirements than Design Parameters, it may be decoupled by the addition of new DPs so as to make the number of FRs and DPs equal to each other.

Theorem 3: Redundant design. When there are more DPs than FRs, the design is either a redundant design or a coupled design.

Theorem 4: Ideal design. In an ideal design, the number of DPs is equal to the number of FRs.

Theorem 5: Path independence of uncoupled design. The information content (design specifications) of uncoupled design is independent of the sequence by which the DPs are changed to satisfy the FRs.

Theorem 6: Equality of information content. All information contents that are relevant to the design task are equally important and no weighting factor should be applied to them.

Theorem 7: Design-manufacturing interface. When the manufacturing or construction system compromises the independence of functional requirements of the product, either the design must be modified or a new manufacturing/ construction process must be developed to maintain the independence of the FRs.

Finally, an observation on the relationship between Axiom 1 and Axiom 2. These axioms are two independent propositions. In an actual design process, one always starts with Axiom 1 and seeks an uncoupled design. Only after several designs that satisfy Axiom 1 are proposed, may one apply Axiom 2 to determine which is the best among those proposed. According to Suh (1990): 'the ability to use Axiom 1 effectively is the hallmark of the creative designer.'

Constraints in design. In addition to the six Design Principles (by which the best design is selected), there are also design constraints which represent the bounds on an acceptable solution. Suh (1990) distinguishes between *input constraints*, which are constraints in design specifications, and *system constraints*, which are imposed by the system in which the design solution must function. The input constraints are usually expressed as limits on size, weight, materials or cost, whereas the system constraints are interfacing bounds such as geometric shape, capacity of machines, and even laws of nature or governmental regulations.

It is sometimes difficult to determine when a certain requirement should be classified as a functional requirement or as a constraint (Suh, 1990). By definition, *a constraint is different from a functional requirement in that a constraint does not have to be independent of other constraints and functional requirements*. In many cases cost is a constraint rather than a functional requirement: its precise value is unimportant, as long as it does not exceed a given limit.

The significance of Suh's design axioms is that they provide principles for distinguishing between good and unacceptable designs and thus put the design process on a scientific basis. Suh's work paved the way for proposing further design principles (see Chapter 4) as well as incorporating them in a specific design methodology for rock engineering.

Example. To illustrate the meaning of FRs and DPs, consider the simple example discussed by Suh (1990) involving the design of a refrigerator door. The first decision that one has to make is what functional requirements the door has to satisfy. Suppose we decided there are two FRs: to provide an insulated enclosure to minimize energy loss and to provide access to the food in the refrigerator.

Given these two FRs, what kind of door would one design? If someone proposed for a DP a vertically hung door that can be opened horizontally, like most commercial refrigerator doors, would this be a good design? If not, why not?

The choice of FRs depends on the way in which the designer hopes to satisfy a set of needs. For the refrigerator door, the energy loss requirement could be eliminated from the list of FRs if the electricity cost of operating the refrigerator is ignored. In this case, the only FR to satisfy is access to the content, so the vertically hung door is perfectly acceptable as a DP and does not violate Axiom 1. However, in an era of high energy costs, the designer's company may not sell many refrigerators if some other company offers a much more energy-efficient system at the same cost. In this situation, the designers did not choose a correct set of FRs when only the FR of access to food was selected.

Clearly, the determination of a good set of FRs from often poorly defined perceived needs requires skill as well as iterations and is an important step in the design process. In fact, one of the major problems in design is that designers try to design intuitively and do not state explicitly the FRs that their design must satisfy.

For the refrigerator door design, we have only considered the energy loss and the access to food in the refrigerator. We were not concerned with the specific ways in which the door was to be hung, or about the insulation technique. Only after one decides whether or not the door should be hung vertically, should one consider the other details. This is an example of an important truism in design: *the FRs and DPs can always be arranged into a hierarchy*, thereby reducing the complexity of the design task immensely.

If the horizontally hung freezer door is accepted as the design solution that satisfies the two FRs, than the DPs of the design solution may now become the constraints at the next level of the FR hierarchy. Therefore, when the FR associated with 'access to the contents' is further relegated into a lower level set of FRs [e.g. FR_1 = we must be able to take out the food in chronological order (i.e. the food that went in first must come out first) and FR_2 = anyone over 1.6m (5 ft 3 in.) tall must be able to reach any item in the refrigerator], the horizontally opening door becomes a constraint at this lower level. The design solution is now locked into concepts that use this type of door.

Now, the question arises: if there are two FRs for the refrigerator door – access to the stored food and minimal energy loss – is the vertically hung door a good design? We can see that the vertically hung door violates Axiom 1, because the two specified FRs are *coupled* by the proposed design. When the door is opened to take out milk, cold air in the refrigerator escapes and gives way to the warm air from outside.

What then is an uncoupled design that somehow does not couple these two FRs? One such uncoupled design of the refrigerator door is the horizontally hinged and vertically opening door used in chest-type freezers. When the door is opened to take out what is inside, the cold air does not escape since cold air is heavier than the warm air. Therefore, this type of chest freezer door does satisfy the first axiom and this DP is an acceptable solution.

Problem. What would be the appropriate FRs and DPs for a bookcase?

Answer
Functional requirements (FRs):
FR_1 = Bookcase to fill the whole wall;
FR_2 = Accommodate books and atlases of different sizes;
FR_3 = Provide for hiding some books from exposure;
FR_4 = Accommodate heavy books and reports;
FR_5 = Provide storage for rolled design drawings.

Constraints
C_1 = Minimal cost;
C_2 = Attractive appearance.
Design parameters (DPs) of the solution:
DP_1 = Specifies the size (height, width and shelf thickness);
DP_2 = Specifies a shorter span for heavy books;
DP_3 = Compartments with doors for books and drawings;
DP_4 = Adjustable shelves for different heights of books;
DP_5 = Material specifications (e.g. cherry wood of high quality).

3.3 GENERAL DESIGN METHODOLOGIES

As defined earlier, design methodology is the collection of procedures and techniques that the designer can use in applying design principles to design. A significant contribution in this respect was made by Pahl and Beitz (1984) in Germany which eventually led to the publication of the VDI-Richtlinie (1987) or standards for engineering design by the Verein Deutscher Ingenieure (Association of German Professional Engineers). These standards or suggested guidelines for design are without parallel anywhere in the world. They provide a systematic approach using an extensive body of well-documented knowledge on the design process accumulated over the years.

On the other hand, Koen (1984) strongly believes that design does not involve nor need a structured methodology; the engineering method is simply the use of engineering heuristics. In between these two extremes, Suh (1990) states that design must involve four distinct aspects of engineering and scientific endeavor: (1) *the problem definition* from a 'fuzzy' array of facts and myths into a coherent statement of the question; (2) *the creative process* of devising a proposed embodiment of solutions; (3) *the analytical process* of determining whether the proposed solution is correct or rational; and (4) *the ultimate check* of the fidelity of the design product to the originally perceived needs.

3.3.1 *Systematic design methodologies*

Pahl and Beitz (1984) have summarized the extensive body of knowledge about systematic design developed in Germany and presented a comprehensive general methodology of engineering design. They split the design process into four main phases: clarification of the task, conceptual design, embodiment design and detail design. The term *embodiment*, in the English translation, means layout design or main design which results in the final arrangement of components, their shapes as well as the materials to be used.

The purpose of Pahl and Beitz (1984) work was to combine the various methods used in Germany and present them in a coherent and practical way. But it was the VDI Guidelines (1987) which constituted a landmark publication of a national standard for design methodology in Germany. Such an undertaking is unprecedented anywhere in the world and sets a challenge for other nations to follow.

The VDI Guidelines 2221 *Systematic Approach to the Design of Technical Systems and Products* is based on the premise that the competitive manufacture and construction of technical systems and products is influenced decisively by the design process. This process is characterized by a wide variety of tasks which must be carried out under conditions specific to each company as well as under those imposed by market trends and technical developments in the state-of-the-art.

The VDI Guidelines 2221 deal with generally valid principles of design independent of a specific branch of industry. They define the design stages and provide the most important principles for design methodology based on the systems approach.

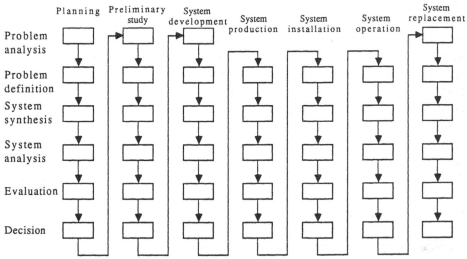

Figure 3.11. Design model of the German systems approach (VDI, 1987).

The systematic approach to design methodology as used in Germany is depicted in Figure 3.11. This approach constitutes a strategy for solving problems and involves such aspects as problem analysis, problem definition, system synthesis, system analysis and evaluation and decision. An iterative approach to problem solving is emphasized and the most effective strategy for design practice consists of subdividing a complex overall problem into defined sub-problems at an early stage. The subsolutions are then combined into an overall solution as shown in Figure 3.12.

The design process is seen as compatible with the different and specific performance requirements of the product to be developed but it also recognizes that the sequence of the design process is additionally determined by general requirements and constraints such as competition, pressures of cost and time, governmental regulations and research findings. It challenges the design engineers to continually update their education and training.

The general approach to design as recommended by the VDI Guidelines 2221 is shown in Figure 3.13. The approach features seven stages with corresponding results. Depending on the task, either all the stages are completed or only some, with other stages being repeated as necessary. In practice, the individual stages are often combined into design phases which assist the overall planning and manage-

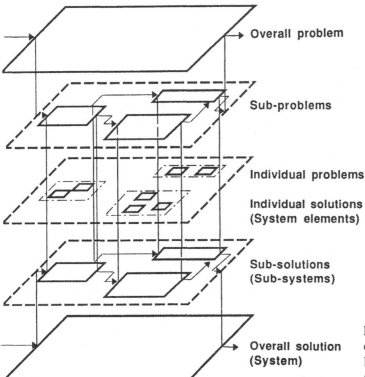

Overall problem

Sub-problems

Individual problems

Individual solutions
(System elements)

Sub-solutions
(Sub-systems)

**Overall solution
(System)**

Figure 3.12. Method of structuring problems and systems (VDI, 1987).

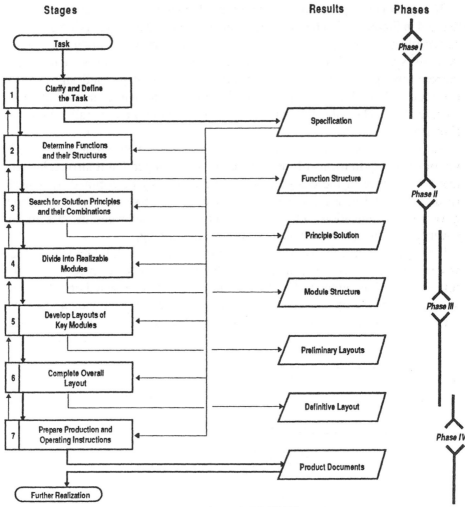

Figure 3.13. Design methodology according to VDI (1987).

ment of the design process. The combination of stages into phases can differ depending on the branch of industry or company. Note that the activities of evaluation and decision are considered necessary in all the stages. These two aspects determine whether or not it is necessary to repeat the preceding stage.

Stage 1, which also starts Phase I (Clarification of the Task), is necessary to clarify and define the requirements of the design task. It includes collecting all the available information. The result is a *specification* (requirement list) – an important working document – which should accompany all subsequent stages, be frequently reviewed and kept up-to-date. After all, important findings in the course of the design process can lead to existing requirements being modified.

Phase II (Conceptual Design) starts with Stage 2 which involves determining the functions: overall function and sub-functions in terms of three aspects: energy, material and signals. The result is one of several function structures which are usually presented as diagrams or descriptions. This leads to Stage 3 in which a search is made for solution principles for all sub-functions and their combination into concept variants. Stage 4 completes the conceptual design phase with the principal solution being divided into realizable modules. A module structure is particularly important in the case of complex products as it facilitates the efficient distribution of the design effort.

Phase III (Embodiment Design) may overlap the previous phase as preliminary layouts of the key modules are developed in Stage 5. This leads to the selection of the best preliminary layout. Next, in Stage 6, these preliminary layouts are finalized by the addition of more detailed information about the assemblies and components previously not included. This results in a definitive layout containing all the essential configuration information for the realization of the product.

In Stage 7, which forms Phase IV (Detail Design), all the final production documents and operating instructions are prepared. This stage may also overlap the previous stage. The result of Stage 7 is documentation in the form of detailed drawings, parts lists as well as testing and operating instructions.

The design process represented by Pahl and Beitz (1984) is somewhat different from that given in Figure 3.13. For comparison, it is depicted in Figure 3.14. In addition, Pahl and Beitz provide more detailed representations of the two individual phases: conceptual design and embodiment design, and they are shown in Figures 3.15 and 3.16.

It is important to note that in the design stages in Figure 3.11, several solution variations will be analyzed and where necessary tested in the form of models or prototypes, and then evaluated. The activities of selecting, optimizing and deciding take place in all stages. In addition, the stages do not have to follow rapidly one after the other but may be carried out iteratively. Furthermore, for some products, it may be necessary to execute the individual stages in parallel. This would apply to software design.

Finally, the decision on the best overall solution is selected after a comparative evaluation of the extent to which the proposed solution variants meet the requirements of the specification.

The work of Pahl and Beitz (1984), as important as it was and one that became the most recognized contribution to design methodology, was not the earliest listing of design stages. This distinction may belong to Rallis (1963) whose work in South Africa did not receive much visibility outside that country but it had a profound influence on the author (Bieniawski, 1984) and resulted in the design methodology depicted in Figure 3.17. It also led to this description of the main stages of a structured design process for rock engineering (Bieniawski, 1988):

1. Recognition of a need or problem.

2. Statement of the problem, including identification of performance objectives and design issues.

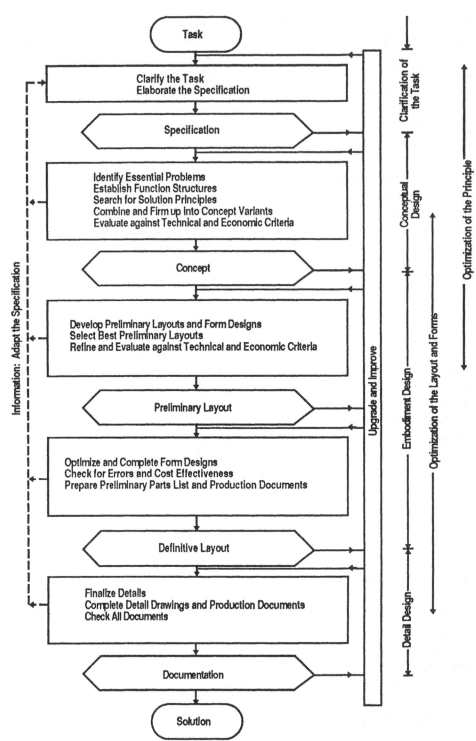

Figure 3.14. Design methodology proposed by Pahl and Beitz (1984).

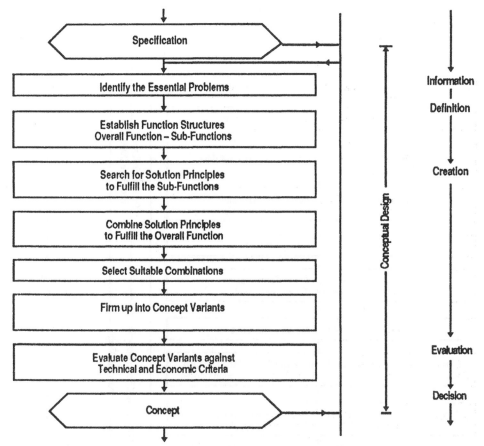

Figure 3.15. Details of steps in conceptual design (Pahl and Beitz, 1984).

3. Collection of information.
4. Concept formulation.
5. Analysis of solution components, including the use of heuristics.
6. Synthesis to create alternative solutions.
7. Evaluation and testing of the solutions.
8. Optimization.
9. Recommendation and communication.
10. Implementation.

Recognition of a need or a problem
The existence of a problem must be recognized before any attempt can be made to solve it. It requires the rather rare ability of asking the right kind of question and calls for a clear recognition of the problem to be solved. In design-type situations, it involves the recognition of a genuine social need, a want or an opportunity.

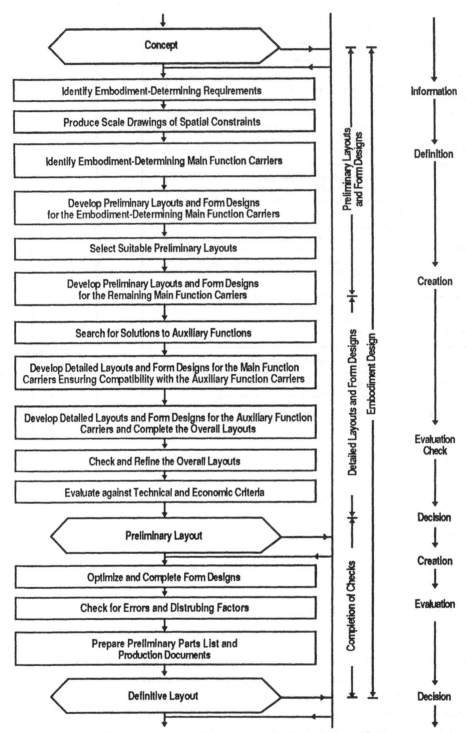

Figure 3.16. Details of steps in embodiment design (Pahl and Beitz, 1984).

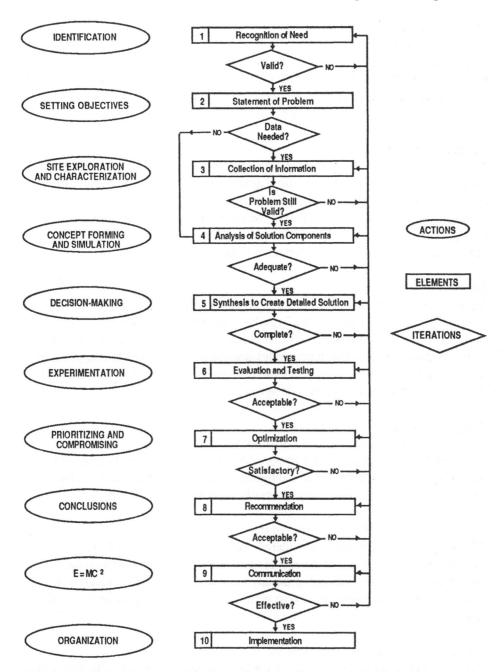

Figure 3.17.General design methodology of Rallis (1963) as adapted for rock engineering (Bieniawski, 1988).

Statement of the problem

Having established that a problem exists, it is then necessary to define it. The ability to define the problem is a most important and difficult task. A poorly formulated problem cannot be expected to produce a good solution. It is at this stage that the design principles discussed earlier are applied, and it is here that independent functional requirements are identified.

Collection of information

This phase involves the gathering, investigation, processing and screening of information to determine the specific characteristics of the problem and to provide the input data for subsequent design analyses. In rock engineering, collection of information includes site exploration, featuring geological and geophysical investigations, laboratory and field testing to establish the characteristics of the rock masses, and evaluation of field stresses and the applied loads.

Concept formulation

Depending on the nature of the problem, either a search is conducted for the most promising method of solution or a hypothesis is selected or invented. Imagination and innovation should be utilized in a manner not unlike that of a creative artist. It is at this stage that the design principles discussed earlier are again applied. It is here that each functional requirement from Stage 2 is met by a corresponding design component representing a design solution.

Analysis of solution components

Design of individual elements is performed as a large design problem is divided into smaller, more manageable components. Design approaches to be used at this stage may involve mathematical and numerical analyses, physical model studies, experiments, and empirical analyses involving the use of heuristics.

Synthesis to create alternative solutions

All the options having been considered and, with the analysis of the individual components completed, the design is directed to creating detailed alternative solutions. This stage should comprise any or all of the following: design specifications, performance predictions, cost estimates, scheduling procedures, and the like.

Evaluation and testing of the solutions

The solutions proposed must now be interpreted and compared with the original functional requirements, design issues, specifications, assumptions, or constraints. This calls for a clear understanding of all the pertinent interacting factors; that is, for the exercise of engineering judgment. The duty of the design engineer is to produce a balanced design involving all the factors which interact.

If, as is frequently the case, such an evaluation shows up deficiencies or suggests more promising alternatives, some or all of the foregoing stages must be repeated. The number of iterations carried out once again calls for judgment, depending on the quality required and the time and money available.

Optimization

Most engineering problems do not have a unique solution. Reconsideration of the solution may thus be necessary in an attempt to approach a feasible compromise between conflicting requirements and resources. The effectiveness of any optimization process depends directly on the clarity with which functional requirements are stated.

Recommendation and communication

Conclusions and recommendations are the outcome of the entire design process. They provide a concise statement of the answer to the problem, point out limitations or restrictions and indicate the direction to be followed in implementing the solution.

The ultimate purpose of the design is the production or construction of a product or project. Achievement of this objective requires the engineer to communicate effectively. Unless the designer can persuade the client or society of the merit of the design it will be stillborn. Effective communication requires that all pertinent facts be properly presented. Hence, the ability to convey thoughts concisely and clearly and to transmit technical knowledge effectively must be acquired by design engineers.*

Implementation

This entails putting the plan into action, and generally involves a high level of organizational skill as well as knowledge and experience of costs, labor, law and equipment. This is the phase which occupies most of the time and energy of construction or manufacturing engineers .

On looking at Figures 3.11, 3.12, and 3.15, the practicing design engineer may well object to the *lack of time*, for going through every one of the many design steps. But it should be borne in mind that:

– most of the steps have to be taken in any case, even if subconsciously, and then unforeseen consequences may arise;

– the deliberate step-by-step procedure ensures that nothing essential has been overlooked or ignored, and is therefore indispensable for original designs;

– in the case of adaptive or improved designs, it is possible to resort to

---

*If a mathematician were to sum up these thoughts, he or she might well do so by the equation (with apologies to Albert Einstein): $E = MC^2$ where E equals effectiveness, M equals mastery of the subject matter, and C equals communication.

time-tested approaches and to reserve the step-by-step procedure for specially promising cases.

– if the designers are expected to produce better results, then they must be given the extra time the systematic approach demands; and

– scheduling is more accurate if a step-by-step method is followed.

3.3.2 *Unstructured use of heuristics*

Such prominent researchers in the field of engineering design as Pahl and Beitz (1984), Yoshikawa (1981), Hubka (1987), and Suh (1990) all favor a systematic design process with some degree of structured or step-by-step features.

Opposing this view is a researcher of no lesser prominence, Billy V. Koen, who believes that engineering design is essentially an unstructured use of engineering heuristics. His formal definition (Koen, 1984) is as follows: *The engineering method is the use of engineering heuristics to cause the best change in a poorly understood situation within the available resources.* A heuristic, according to Koen, is anything that provides a plausible aid or direction in the solution of a problem but is in the final analysis unjustified, incapable of justification and fallible. It is anything that is used to guide, discover and reveal – possibly but not necessarily – a correct way to solve a problem. A heuristic has four characteristics: it does not guarantee a solution, it may contradict other heuristics; it reduces the search time for solving a problem; and its acceptability depends on the immediate context instead of on an absolute standard. In simple terms, a heuristics is a 'rule of thumb.'

Deriving from a Greek word, 'heuristic' means *serving to discover.* Some historians attribute the earliest mention of the concept of heuristics to Socrates in about 469 B.C. and others identify it with the mathematician, Pappus, around 300 A.D.

Although no taxonomy of heuristics is as yet available, some types of engineering heuristics are:

1. Simple rules of thumb and orders of magnitude,
2. Factors of safety,
3. Heuristics that determine the engineer's attitude towards work,
4. Heuristics that engineers use to keep risk within acceptable bounds, and
5. Rules of thumb that are important in resource allocation.

Examples are:

– *The yield strength of a material is equal to a 0.02% offset on the stress-strain curve.*

– *A properly designed bolt should have at least one and one-half turns in the threads.*

– *Always give an answer.*

– *Use a morphology to solve an engineering design problem.*

A typical set of heuristics is found in expert system programming of rules for artifical intelligence concepts. In rock engineering, a heuristic would be a rule that the vertical stress in the ground increases at a rate of 1 psi per ft of depth or that the bolt length should be one-half the roof span in civil engineering tunnels. The value of a factor of safety considered acceptable for a pillar or a rock slope is also a heuristic.

Koen (1985) submits that structured methodologies are inadequate as a definition of design for four reasons: (1) they resemble the eccentric vision of their author; (2) engineers cannot simply work their way down a list of steps, they must circulate freely within the proposed plan (iterating); (3) none of the structured design processes so far proposed recognizes the full spectrum of heuristics essential to a proper definition of the engineering method; and (4) in actual practice it is highly unlikely that engineers follow any structure proposed to explain their work.

Arising out of his definition of engineering, Koen (1984) proposed just one principal rule for implementing the engineering method: *Choose the best heuristics for use in every instance from what your personal state-of-the-art* ('sota') *takes to be the sota representing best engineering practice at the time you are required to choose.*

Koen's views have not gone unchallenged! Andrews (1987) pointed out that any adequate definition of the engineering method must also recognize that the successful solution of an engineering problem requires mastery of the relevant scientific principles and related technology, as well as the ability to apply this information in both routine and creative ways. Hence, not only experience but also creative ability must be regarded as essential ingredients in the successful implementation of the engineering method. He also challenged Koen's definition of a heuristic particularly that it is *unjustified* and *incapable of justification*. A better definition would be: *A heuristic is any plausible but fallible solution aid when it is considered for implementation.*

Andrews (1987) also takes issue with the four reasons given by Koen (1985) as to why a systematic approach to engineering design would not work. After counteracting and dismissing all four reasons, Koen's definition of the engineering method as simply *the ad hoc use of heuristics* is seen as too simplistic. Instead, this definition is recommended: *The engineering method is a logical, ordered and systematic procedure or plan for solving engineering problems in an effective and efficient manner which incorporates the use of engineering heuristics and heuristic reasoning.*

This view has also been supported by Hazelrigg (1988) who cautioned, however, that both Andrews and Koen seem to confuse engineering with science. He objected to an assumption that engineering is 'problem-solving.' It is important to understand the correct definition that *engineering design is a process of decision making.* The results of problem-solving exercises provide information helpful to decision-making, but decision-making is a human process, not a process of the physical world described by the laws of nature.

Devon (1988) pointed out that Koen's claim that *all is heuristic* is a principle that can be taken too far: an engineer's rule-of-thumb is not on a par with Newton's *Principia Mathematica* or the US Constitution. Differences in sophistication and significance should not be blurred.

It is obvious from the above discussion that the systematic engineering design process is not only the most effective approach but also one that is flexible enough to incorporate the use of engineering heuristics and heuristic reasoning.

3.4 IN SEARCH OF DESIGN SCIENCE

The NSF funded study (ASME, 1986) identified *design theory and methodology* as a new engineering discipline and pointed out that this new field has some, but not all, of the attributes of an engineering science, as well as some attributes of creative art. It is also clear from the preceding discussion that much scientific research has been directed to design theory as well as to design methodology.

Can one therefore claim that *design science* has come into being? For some, design science may imply a contradiction because design could be seen as an intuitive, mental occurrence that should defy description. For others, design science is as readily acceptable as is the proliferation of such names as: political science, library science, sports science, and even domestic science.

The Webster Dictionary defines science as (Guralnik, 1986): (a) from Latin *scientia*, having knowledge; (b) a branch of study concerned with observation and classification of facts, principles and methods; (c) accumulated systematized knowledge. Moreover, a science must pass through various stages of maturity, namely: (1) description of phenomena; (2) categorization in terms of apparently significant concepts; (3) ordered pattern which may be deemed a model; (4) isolation and test of phenomena with implied reproducibility by independent observers; and (5) quantification.

Based on these considerations, there clearly exists a co-ordinated body of knowledge ABOUT designing that can be termed *design science* (Gregory, 1986; Willem, 1990). Nevertheless, it may be noted that in the case of design science, as in many other fields of science, empiricism and technological needs have always preceded the development of the science *per se*, e.g. thermodynanics evolved only after the development of the steam engine.

The scope of design science includes two factors which are essential for understanding engineering design: they are (as shown in Figure 3.8) the knowledge about designed artifacts (technical systems) and the knowledge about the process of designing them. According to Hubka and Eder (1988), existing knowledge about engineering design, that is, *design science*, which considers the design products within the social and economic context, is devoted to such questions as:

– how can technical processes or systems be modelled?

– how can we find out how good a design product will be for its purpose *before it is built?*

– how can technical processes or systems be classified to characterize them?

– how can we describe a stepwise (systematic, methodical, complete, ideal) design process so that a practical designer can select a useful way for a real problem?

Design science is a reality as a discipline, but it must be co-ordinated with existing technical knowledge including the ideas and concepts of general systems theory (Klir, 1969). However, design science alone cannot make a potential engineer into a good designer. Neither can design practice on its own yield an effective way of learning how to design, nor is an extensive knowledge of hard sciences and technology solely adequate: all three aspects have to be brought under one umbrella. The technical knowledge must be co-ordinated with design theory and methodology and adequate practical experience must also be provided.

According to Simon (1969), *design science is concerned with how things ought to be, natural sciences – on the other hand – are concerned with how things are.* However, as seen by Willem (1990), the principal function of design is to *have an effect.* It is through design that science exceeds being pure knowledge and participates in creating effect. *Design science* is a forum for bringing together *design* and *science.*

CHAPTER 4

Design theory for rock engineering

Engineering design is a crucial component of the industrial product realization process. Over 70 percent of the life cycle cost of a product is determined during design.

National Research Council (1991)

What is the role of theory in rock engineering? *One only has to contemplate the complexity of geologic structures in the field to appreciate why the development of a sound theoretical basis for engineering design in rock has been slow compared to progress in the design of structures involving only man-made materials; design in rock remains 'more art than science' and rules, where they exist, tend to be empirical; this situation must change,* wrote C. Fairhurst (1976).

One important use of theory in rock engineering is that it permits an examination of various alternative design and construction procedures and gives the basis for at least a tentative design. Another use is to provide a basis for comparing the anticipated conditions with the encountered conditions.

Thus, the role of theory is clear: it permits the development of initial concepts of design on the basis of a range of reasonable assumptions concerning *in situ* rock properties and the state of stress. It also provides essential guidance on design methodology. Finally, it furnishes a norm against which the results of the field observations can be compared as a basis for assessing the actual conditions.

Design theory is needed in rock engineering as a theory *about* designing, just as the above discussed theory *for* designing forms an essential foundation to the designer's pool of technical knowledge. Design theory in rock engineering is directed to providing important general *design principles* that can serve as valuable guides to design in the geologic environment.

Before discussing design theory and methodology for rock engineering, it is important to recognize that designing a mine or a tunnel is different from designing a conventional structure, a building or a bridge.

In a conventional engineering design, the external loads to be applied are first determined and a material is then prescribed with the appropriate strength and deformation characteristics, following which the structural geometry is selected. In rock engineering, the designer deals with complex rock masses and specific material properties cannot be prescribed to meet design requirements. Furthermore, the applied loads are not as important in rock masses as the forces resulting from the redistribution of the original stresses, i.e. those existing before the excavation was made. Also, a number of possible failure modes can exist in a rock structure so that determination of the 'material strength' is a major problem. Finally, the geometry of a structure in rock may depend on the configuration of the geological features. Hence, the design of an excavation in rock must include a thorough appraisal of the geological conditions and especially of possible geological hazards.

Does this mean that the engineering design process as discussed in the previous section cannot be applied in rock engineering design? Certainly not! It does mean, however, that the design of excavations in rock requires extra considerations of the special geotechnical conditions.

In essence, rock engineering design incorporates planning the location of structures, determining their dimensions, shapes, orientations, layout, excavation procedures (blasting or machine boring), support selection and instrumentation. The rock engineer studies the original in situ stresses, monitors the changes in stress due to mining or tunneling, determines rock properties, analyzes stresses, deformations and water conditions, and interprets instrumentation data.

4.1 DESIGN APPROACHES IN ROCK ENGINEERING

As stated earlier, the application of design methodology to rock engineering has not received as much attention as in other engineering fields. The result has been excessive safety factors in many civil and mining engineering projects. Moreover, while extensive research is being conducted in rock engineering today, there still seems to be a major problem in 'translating' the research findings into innovative and concise design procedures.

The design methods which are currently used in rock engineering can be categorized as follows:
a) Analytical methods;
b) Observational methods;
c) Empirical methods.

Analytical methods: Utilize the analyses of stresses and deformations around excavations. They include such techniques as closed form solutions, numerical methods (finite element, finite difference, boundary element), analog simulations (electrical and photoelastic) and physical modeling.

Observational methods: Rely on actual monitoring of ground movement during excavation to detect measurable instability, and on the analysis of ground-support interaction. Although considered as separate methods, observational approaches are the only way to check the results and predictions of the other methods.

Empirical methods: Frequently used in rock engineering practice. They assess the stability of mines and tunnels by the use of statistical analyses of underground observations. Engineering rock mass classifications are the best known empirical approach for assessing the stability of excavations in rock. They have received increasing attention in recent years and in many projects this approach has been utilized as the only practical basis for design.

All the above methods require geological input and consideration of statutory safety regulations.

Early guidelines for a design procedure in rock engineering were provided over a decade ago by Hoek and Brown (1980) who suggested a number of steps for underground excavation design. They emphasized that the basic aim of any underground excavation design should be to utilize the rock mass itself as the principal structural medium.

In a text specifically devoted to rock mechanics design, Bieniawski (1984) presented design charts for applications in mining and tunneling, which included a design chart for mining excavations and a series of flow charts for the design of tunnels and chambers in rock. Subsequent work (Bieniawski, 1988) also included a discussion of the possible design principles for evaluation of alternative designs.

Brown (1985) emphasized the need for improved design as a bridge between theory and practice in rock engineering. He listed these components of a generalized rock engineering program for the design of civil and mining engineering projects:

1. Site characterization (definition of geomechanical properties of the host rock mass).

2. Geotechnical model formulation (conceptualization of site characterization data).

3. Design analysis (selection and application of mathematical and computational schemes for study of trial designs).

4. Rock mass performance monitoring (measurement of the performance of the host rock mass during and after excavation).

5. Retrospective analysis (quantification of in-situ rock mass properties and identification of dominant modes of rock mass response).

Brown (1985) emphasized the use of theory and analysis in underground excavation design. Theory, based on the principles of mechanics, should be used in the design analysis stage, in the design of monitoring programs, in the interpretation of monitoring results and in retrospective analysis and redesign. To assist with design analyses, Starfield and Cundall (1989) emphasized the need for a methodology of rock mechanics modeling.

4.2 DESIGN PRINCIPLES FOR ROCK ENGINEERING

Using the approach advocated by Suh (1990), six design axioms are proposed for rock engineering as the principles for evaluating and optimizing alternative designs. While Suh's two original axioms (see 3.2.3) are considered *necessary*, they are *not sufficient* for rock engineering design because the behavior of rock masses is governed by the geologic environment which imposes unique constraints not found in other branches of engineering.

Six principles of design are proposed for rock engineering:

1. *Independence Principle*: There exists a minimum set of independent functional requirements that completely characterize the design objectives for a specific need.

2. *Minimum Uncertainty Principle*: The best design is one which poses the least uncertainity concerning geologic conditions.

3. *Simplicity Principle*: The complexity of any design solution can be minimized by creating the fewest number of design components forming a part of the design solution and each corresponding to the appropriate functional requirement. In this way, the design objectives are uniquely satisified in terms of the problem definition.*

4. *State-of-the-art Principle*: The best design maximizes technology transfer of state-of-the-art research findings and the current best practices.

5. *Optimization Principle*: The best design is the *optimal design*, evolved from quantitative evaluation of alternative designs based on the optimization theory, including cost effectiveness considerations.

6. *Constructibility Principle*: The best design facilitates the most efficient construction of the rock engineering structure by enabling the most appropriate construction method and sequence, and a fair construction contract.

4.2.1 *Independence Principle*

Justification of these design principles must start by emphasizing that one of the major problems in design is that design objectives are often ill-defined and designers do not state explicitly the functional requirements their design must satisfy. They often try to design intuitively and do not recognize the need to reiterate the functional requirements until a satisfactory design results. When a new set of functional requirements are established, the corresponding solution may be completely different from those previously tried.

Thus, proper problem definition is most important in design and Design Principle 1 is directed to that purpose.

Since the designer can arbitrarily define the functional requirements to meet a

---

*Note that the term *design components* is used here to characterize a design solution instead of Suh's *design parameters*. In rock engineering, *parameters* could be a misleading word because the term is so strongly associated with site investigations (e.g. geological parameters).

certain perceived need, an acceptable set of functional requirements is not necessarily unique. Moreover, corresponding to a set of functional requirements there can be many design solutions. This then provides ample scope for creativity and produces design winners and losers.

In summary, Design Principle 1 states: *Maintain the minimum of independent functional requirements which completely characterize the design objectives for a specific need.*

This principle has two corollaries:

Corollary 1.1: FRs have hierarchies and they can be relegated to lower levels.

Corollary 1.2: FRs at one level cannot be relegated into the next level of the FR hierarchy without first going over to the physical domain and developing a solution that satisfies the first level FRs.

4.2.2 *Minimum Uncertainty Principle*

This principle is proposed for rock engineering because unlike other engineering materials rock and rock masses cannot be fully characterized for engineering design in a manner that steel or concrete can. Moreover, rock masses are complex geologic structures governed by large scale geologic discontinuities and are difficult to test as a full scale prototype. Accordingly, extrapolation of data from small-scale laboratory samples to large-scale *in situ* features will always involve a degree of uncertainty. In fact, the questions: 'When do we determine that the needs for site characterization are met?' and 'How much information is enough?' have no consensus of answers in the rock engineering community.

Moreover, it is quite common that only limited information is available on ground conditions at the time a rock engineering project is designed. Mechanical characteristics and even boundaries between strata remain uncertain, and important faults or cavities may have been missed even by a thorough investigation. This leads to reliance on monitoring the behavior of rock masses during construction and even to design adjustments during construction as advocated by the New Austrian Tunneling Method (Sauer, 1988).

Given the uncertainties and complexities in the processes affecting rock engineering design, Principle 2 is proposed:

The best design is one which poses the least uncertainity concerning geologic conditions.

This principle has two corollaries which are self-evident:

Corollary 2.1: Any data from site investigation must be directly traceable to the design objectives represented by the functional requirements (FRs).

Corollary 2.2: Any data from site characterization must be used in the design solution represented by the design components (DCs).

4.2.3 *Simplicity Principle*

It is the role of Design Principle 3 to assist in deciding which design is the best.

First, a good designer would satisfy Principle 1 by choosing a minimum number of functional requirements. Since, however, the output of the design process is in the form of drawings, specifications, and other relevant knowledge required to create the physical entity, the best design solution should be as simple as possible, so the design output can be produced with minimal effort. This is the essence of Principle 3, called the Simplicity Principle. Its motto is: *the simpler, the better.*

This principle is Suh's Axiom 2 which states (Suh, 1990): 'Among the designs that satisfy the condition of the independence of functional requirements, the one with the minimum complexity is the best design.' The term 'best' is used in a relative sense because there are potentially an infinite number of designs that can satisfy a given set of functional requirements. These designs are distinguished by their own characteristic solutions featuring design components (DCs) which fulfill the appropriate functional requirements.

In summary, Design Principle 3 states: *Minimize the complexity of the design solution by creating a minimum number of design components corresponding to each functional requirement.*

This principle has three corollaries:

Corollary 3.1: The DCs and FRs are related in such a way that a specific DC can be adjusted to satisfy its corresponding FR without affecting other functional requirements.

Corollary 3.2: DCs have hierarchies and they can be relegated to lower levels.

Corollary 3.3: FRs at one level cannot be relegated into the next level without first going over to the physical domain and developing DCs that satisfy the first level FRs. Thus, one has to travel back and forth between the functional domain and the physical domain in developing the FR and the DC hierarchy.

4.2.4 *State-of-the-art Principle*

In spite of extensive research performed in the field of rock engineering since the First International Congress in Rock Mechanics held in Lisbon in 1966, innovation in rock engineering design has not proceeded as rapidly as in other engineering fields. This is mainly due to industry's cautious reaction to change and reluctance to introduce new products and approaches until they have been proven elsewhere. Innovation is often a response to a sudden emergency with no conventional solution possible.

It is interesting to note that rock bolts and shotcrete represented the last major innovations in mine roof support technologies. But, while rock bolts were enormously successful since their introduction around 1940s, even today rock bolt parameters and layout are specified primarily on the basis of empirical procedures and practical experience. Similarly, a conference on rock bursts in 1985 showed that scientific knowledge existed more than 10 years earlier for controlling and reducing rock burst hazards in mines and tunnels (Brown, 1985).

It is submitted that great strides have been made in rock engineering research which must be incorporated in rock engineering practice through innovative designs featuring state-of-the-art technology.

Design Principle 4 states: *The best design maximizes technology transfer of the state-of-the-art findings and the current best practices.*

4.2.5 *Optimization Principle*

Not too long ago, some design approaches were characterized picturesquely as 'quick and dirty.' Quick meant easy to calculate; dirty, approximate. These procedures provided conservative design in the sense of being workable but not necessarily economical. More recent formulations of design as an optimization problem led to the use of computer methods which enable sophisticated modeling and fast computation. It is believed that a good design is one achieved by the use of optimization techniques resulting in a quick and innovative product.

An important contribution in this respect was provided by Wilde (1978) and by Siddall (1982) who defined *optimal design* of a device as the feasible plan that makes it as good as possible according to some quantitative measure of effectiveness. In essence, optimization is seen as the foremost goal of any engineering design and optimal design is viewed as applying the optimization theory to engineering design. The reference (Siddall, 1982) contains full details of the optimization theory, including computer programs, for bringing optimal design into everyday design practice.

Optimization is crucial in design because most engineering problems do not have a unique solution. Reconsideration of the solution may be necessary in an attempt to approach a feasible compromise between the often conflicting requirements and resources.

Design Principle 5 states: *The best design is an optimal design which is evolved from the quantitative evaluation of alternative designs based on optimization theory, including cost effectiveness considerations.*

4.2.6 *Constructibility Principle*

In rock engineering design, we can envision three domains: functional domain, physical domain and construction domain. Each of these domains is defined by multiparameters or multivariables. During the design stage, the functional requirements (FRs) must be satisfied by choosing a proper set of design components (DCs), whereas during the construction phase, the DCs must be satisfied by selecting an optimum set of construction procedures (CPs). Effective design for constructibility requires the optimization of relationships among the functional, design and construction domains; so, there is a relationship between functional requirements and construction procedures.

Design Principle 6 states: *The best design facilitates most efficient construction of the rock engineering structure with the components of the design solution being implemented by the most efficient construction procedures.*

4.3 DESIGN METHODOLOGY FOR ROCK ENGINEERING

It is submitted that design methodology for rock engineering would benefit from a structured process featuring a number of design stages but one that would not constitute a 'straightjacket'; rather it should be a flexible framework adaptable to the problem at hand. One should thus visualize design methodology as a checklist (not unlike the one used by pilots before take off) or a road-map guiding the designer to the fulfillment of the problem objectives by evolving the best design option. It is thus a sequence of steps or activities within which a design can unfold logically. It serves as a useful reference of where we are, where we ought to be, and what the next step should be within the overall work plan.

Accordingly, the writer believes that an effective design methodology for rock engineering can include elements of a systematic design process, such as developed by Pahl and Beitz (1984) and by Bieniawski (1988), and can also incorporate the use of engineering heuristics, as propounded by Koen (1984).

A comprehensive design methodology is not just a sequence of flow charts for step-by-step design; this has been done before (Hoek and Brown, 1980; Bieniawski, 1984). To be comprehensive, a design methodology must incorporate design principles which can be used to evaluate designs and to select the optimum one fulfilling the perceived objectives. A design methodology must indeed recommend an order of design stages but these must be so structured as to assist in effective decision making and promote design innovation in accordance with the design principles.

The concept. The proposed design methodology is seen as a systematic decision-making process aimed at satisfying the perceived needs, identified by independent functional requirements. Creative design solutions are represented by design components which meet the corresponding functional requirements and facilitate selection of efficient construction procedures. There are thus three broad domains involved:

Domain of Objectives	→	Domain of Design Solutions	→	Domain of Construction
FRs	*Creative process*	DCs	*Optimization process*	CPs
Functional Requirements	→	Design Components	→	Construction Procedures

The methodology. Summarized in Figure 4.1 are the main features of the proposed Design Methodology for Rock Engineering. The design principles indicated ensure that a good design is produced and offer a basis for comparing and selecting designs.

The design process starts with the *Statement of the problem* which recognizes a societal need. This must provide a clear and concise narrative statement followed by itemized deposition of performance objectives and the ensuing design issues.

Identification of *Functional requirements and constraints* constitutes the next step which formalizes the need consistent with the design principles presented in Section 4.2. This means that Design Principle 1 must be satisfied here by specifying the smallest number of *independent* funtional requirements (FRs) which completely characterize the design objectives for the need in question. Constraints must also be clearly stated but they may be *interdependent*. Next, the design issues are reviewed and the design variables, arising from these issues, identified. The designer may now proceed to select functional requirements which define the design problem.

Collection of information is the third stage in the design methodology and must include a thorough geotechnical site exploration compatible with the design issues. Guidelines and procedures are available for this purpose (Hoek, 1987; Salamon, 1988) but care must be exercised that the information to be collected satisfies Design Principle 2 (Minimum Uncertainty) and its two corollaries (see Section 4.2.2).

Concept formulation, analysis of solution components, and synthesis into alternative solutions are the three subsequent stages where design analyses are performed and where creative solutions are sought. All these stages must satisfy Design Principle 3 by which the design components, which represent the elements of the design solution, must satisfy explicitly the functional requirements in Stage 2.

It is at the stage of *concept formulation* that the designer's ingenuity and imagination can be realized to the fullest. It is an opportunity to seek truly innovative alternative solutions. This stage challenges the designer to be creative but, as a minimum, the designer must be up-to-date with available technology. For this purpose, Design Principle 4 must be observed. This is the State-of-the-art Principle which assures that, during the *analysis of solution components and the synthesis to create alternative solutions*, the latest research findings are incorporated as technology transfer into design practice. The latest techniques representing analytical, observational, and empirical methods should be utilized. Most of all, it is here that the use of *heuristics* will be most appropriate together with any expert systems available to assist with the problem at hand. Where needed, the actual state-of-the-art should be advanced by new ideas leading to new solutions.

The State-of-the-art Principle is very important. Koen (1984) even argued that it is the collection of state of the art heuristics known to the designer that defines the very nature of engineering. It is obvious that depending on how up to date the

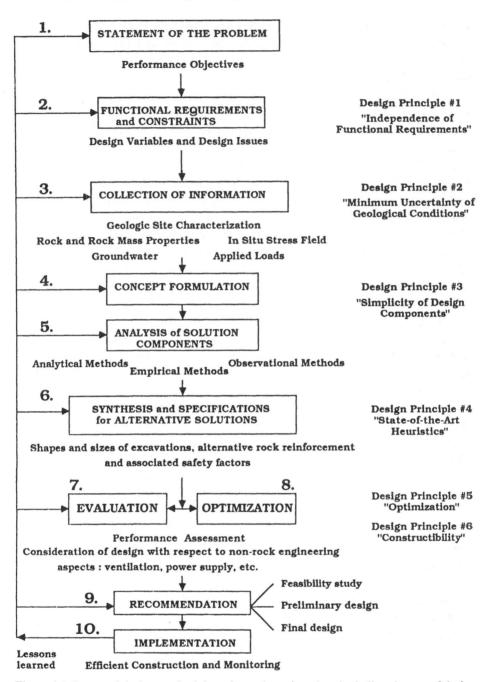

Figure 4.1. Proposed design methodology for rock engineering, including the use of design principles.

designer is, one's design can be far superior to that of a competing person.

Specifications for alternative solutions should include shapes and sizes of excavations, alternative rock reinforcement, and associated safety factors. Moreover, these specifications should be supported by engineering drawings which, in these days of CAD software abundance, are not a major undertaking but one that greatly facilitates an understanding of the proposed solution.

An important consideration at this stage should also be how the design will be realized in practice, that is, by what construction procedures. It is here that Design Principle 6 must be first observed. This mandates that the design components of the design solution can be satisfied by appropriate construction procedures, with maximum efficiency.

The design methodology now moves to Stage 7: *Evaluation*. Not unlike the film editor who chooses the best movie version from various alternative 'cuts' or scenes, the designer must now evaluate the alternative solutions; firstly, to ensure that they fulfill the functional requirements and hence meet the design objectives, and secondly, to select the preferred solution. This evaluation must include reviewing all six Design Principles and constraints. In addition, if not already included as a constraint, cost estimates for each solution must be considered. Moreover, sometimes design objectives might prescribe *necessary* features as well as *desirable* features (which would be welcome but are not essential). Pahl and Beitz (1984) use this 'desirable' feature as a criterion for evaluation of designs when a number of solutions have provided the 'essential' features.

The evaluation stage does not have to lead to a single design choice. Two solutions or even three solutions might be selected from typically half-a-dozen alternatives, each with special technical features but also different costs. This sets the stage for the next step: optimization.

Optimization, considered the foremost goal of any engineering design, is the stage where Design Principle 5 must be fulfilled utilizing the optimization theory (Wilde, 1978) and associated techniques (Siddall, 1982). The design solution eventually selected must again satisfy all the Design Principles, including the *Design for Constructibility* Principle 6. Based on this approach, the final solution emerges which must be recommended for acceptance by the client.

The proposed Design Methodology for Rock Engineering specifies stage 9: *Recommendation* as a separate step because of the importance of effective communication by engineers. At this stage, a comprehensive design report has to be prepared including the description of the special features of the proposed design and all its merits. Unless these descriptions are clearly presented even the best technical design may not win acceptance by the sponsor.

Included in the descriptions and specifications for the recommended design should be a document entitled *Geotechnical Design Summary and Geotechnical*

Data Interpretations which should be included as part of the contractual documents, leading to fair risk-sharing between the project owner and the contractor.

Implementation is the tenth and last stage of the proposed design methodology. It is here that the construction of the project takes place and the design itself is realized. While this stage is important for all fields of engineering, it has a special significance in rock engineering due to the unique uncertainty of geological conditions. The behavior of rock excavations must often be monitored during construction with possible adjustments to the final design being introduced and even changes made to design objectives. In effect, in rock engineering, the design is not complete until construction is completed.

Furthermore, again due to these uncertainties in rock mass conditions, any lessons learned during construction, as to the suitability of any design assumptions, models or predictions used, should be recorded to serve as a future data base for designers. This may even result in the development of new state-of-the-art heuristics.

In summary, the proposed design methodology for rock engineering consists of ten steps and features six design principles enabling comparison of designs and selection of the best design. Most of all, the design process is seen as involving three principal domains:

1. Objectives Domain, where functional requirements are specified;
2. Design Solutions Domain, where design components are invented to meet the functional requirements; and
3. Construction Domain, where construction procedures implement the design solution components.

Linking these three domains are two processes: the *creative process* necessary to move from the domain of objectives to the domain of design solutions, and the *optimization process* necessary to move from the domain of design solutions to the domain of construction. As a result of this design methodology, the final construction procedures will fulfill the original functional requirements thus satisfying the perceived need and resulting in a *good* design. The design principles provide a yardstick for design quality as well as design selection, while the design methodology not only incorporates these principles but also provides a flexible roadmap for moving from a clearly stated objective to the completed construction of an innovative design.

4.4 EXAMPLE

A Design Mini-Project
The purpose of this example is to demonstrate the selection of the functional requirements, constraints and design components.

Objective. Design a rock bolt to provide effective roof support for long-term applications in underground coal mines.

Example 75

Functional requirements:

   FR_1 = ability to resist corrosion (long-term use)

   FR_2 = anchorability in weak rock conditions

   FR_3 = effectiveness over full bolt length.

Constraints:

   C_1 = fast, easy installation

   C_2 = permissibility in coal mines (licenseable for 'fiery' conditions)

   C_3 = cost comparable to present day roof bolts.

Design concept. Using the analogies of a bicycle tube and a 'flat jack' (hydraulic loading device), a concept is selected whereby a hollow steel tube is inserted in the borehole and expanded hydraulically by pressure within the tube. The expanded tube produces the radial stress neccessary to reinforce the rock. The bolt is initially 'collapsed', i.e. flattened and rolled into a smaller diameter.

Design components:

   DC_1 = expandable hollow steel tube

   DC_2 = high strength steel

   DC_3 = radial pressure of 4,000 psi (27.5 MPa) to expand a collapsed steel tube and to provide the necessary radial stress (pumps commercially available).

   Observing the constraints: C_1 and C_2 are fulfilled, C_3 must be justified, i.e. the cost of an installed bolt (including material, transport, storage, drilling and installation). The following information was determined for comparison (Kicker, 1990):

Bolt type	Cost/unit	Effectiveness*
mechanically tensioned bolt	$19.29	54
grouted resin bolt	$20.28	74
'split-set' friction bolt	$17.47	62
hydraulically expanded bolt	$16.79	92

Implementation. This hydraulically expanded rock bolt has been designed, manufactured and marketed by Atlas Copco of Sweden under the trade name 'Swellex' bolt.

*Represents the overall rock support quality on a scale of 100 in terms of the design objective of this example [derived from individual ratings of 1 to 10 in ten categories: reliability (design capacity, active support, long-term usage); failure type versus support (six sub-categories); and ability for accurate non-destructive testing].

CHAPTER 5

Design education for rock engineering

I am struck by the lack of conversation about what pedagogy means, and what makes it sucessful. It is our profession, yet it is mysteriously absent from our professional discourse... We do not argue enough about what ought to be the common intellectual property of educated men and women.

Donald Kennedy, Stanford President (April 1990)

If the lack of a comprehensive design methodology for rock engineering has hampered innovative design in this field, an even greater case can be made for the unsatisfactory state of design education. This is currently a major concern in most engineering disciplines in the USA and elsewhere. Thus, a discourse on engineering design must include a discussion of design education. In any case, it may already be too late to change the habits of many practicing rock engineers, so that our attention must be directed to the designers of the future.

The writer believes that not enough emphasis is placed on engineering design in the undergraduate and graduate courses for engineers. As a result, engineerng students graduate from universities without an incentive to strive for innovative engineering designs in their practice.

As a mitigation for the engineering professors, the teaching of design is a demanding and particularly challenging task. There are three primary reasons why design is sometimes taught poorly or not at all: (1) professors are better at analysis than at design; (2) design is difficult to teach, teaching analysis is easier; (3) in spite of all the discussion about design, it is not considered in some quarters of academia to be very important in terms of its 'scientific value' for tenure review. Most of all, however, very few engineering professors have sufficient design experience to teach this subject. After all, the required educational qualification for a person entering the academic profession is a Ph. D. degree – hence, research and not practical experience.

Nevertheless, the American Society for Engineering Education, through its Design in Engineering Education Division has consistently emphasized the need for better engineering design education, and many universities have responded. Departments of Engineering Design now exist at such universities as Stanford, MIT, Tufts and Cambridge (England) while Harvard has a Graduate School of Design. Moreover, Carnegie-Mellon (Pittsburgh), and Texas A & M Universities have established Engineering Design Research Centers. In addition, specific professorships in design engineering have been endowed at many universities. However, none of this is taking place in the area of rock engineering!

The following observations are derived from teaching engineering design in Europe and America (McMahon, 1989):

1. Design requires a different teaching approach.
2. Design is best learned by doing.
3. A clear methodology is important.
4. Design should be integrated with analysis courses, and
5. Design should be presented as a transformation from the functional description to a physical entity.

It is also appropriate to state some observations concerning the difficulties which students have with design. These are:

1. Grappling with the statement of the problem and the need to define the objectives clearly.
2. Realizing there is no 'correct' solution.
3. Lack of appreciation of iterative techniques.
4. Developing the skills for decision making.
5. Paying attention to details and to preparing clear drawings.
6. Communication.

The question, therefore, is: 'How would one *design* the engineering graduate who would be good at design in the industrial environment?' At the start, however, there is a problem : there are too many subjects to fit into the curriculum. This is particularly constraining in America where 'general education' courses carve out one-third of the total credits needed for graduation during the student's four years at the university. In other countries, these general education requirements are satisfied by a comprehensive high school education! Moreover, engineering education in Europe is generally a five-year program.

So what, can we do to improve a four-year curriculum? In essence, because we cannot teach the student everything about the discipline, we should:

1. Place emphasis on the basic tools and techniques rather than specific information; i.e. teach how to apply knowledge.
2. Make design methodology and problem-solving techniques a part of the curriculum. Design methodology and nomenclature should be uniform throughout the curriculum.
3. Initiate realistic design projects, including construction, as a part of the curriculum.
4. Include group projects which encourage teamwork and communication. There should be a place for interdisciplinary efforts.
5. Use case studies in most courses to teach problem evaluation and the application of engineering science to practice.
6. Incorporate information gathering in many courses (literature searches, use of handbooks, expert systems, etc).

7. Encourage students to explore specific technologies and applications of technologies on an independent basis; i.e. self-education.

8. Arrange for design solutions by the students to be presented to a design jury made up of practicing engineers.

It is believed that engineering design cannot be taught in only *design* courses, and particularly not in only *capstone design* courses. The basics of problem definition, analysis and execution of specific solutions (with applications) belong to all courses. However, design methodology is a crucial element of the curriculum and deserves special attention. This is so because we must recognize that engineering design is the essence of engineering education.

Concerning design education in *rock engineering*, the ideal approach is that adopted in Germany, as described by Rauhut (1989). There, practical experience goes hand in hand with the academic training being received at the university. A student who intends to become a mining engineer must work for one year in industry after finishing high school and before starting university! This idea is achievable in Germany due to the committment of the German mining industry to this requirement and guaranteeing employment for a pre-university year to all aspiring mining students. Moreover, the German students are also encouraged to obtain further practical experience during the course of their studies, taking time out to do it. As for their academic skills, the German students must acquire 'exceptional education in all the basic fundamentals of engineering' and 'new and imaginative solutions requiring innovative and creative thought' are often emphasized as 'the only possibility' (Rauhut, 1989). It is not surprising that with such credentials German design engineers excel at their jobs and set the standard for the rest of the world!

As for an actual curriculum of engineering design which might be adaptable to rock engineering, the interested reader should study the approach used at the Tufts University Department of Engineering Design (Crochetiere, 1987). Tufts offers M.S. and Ph.D. degrees specifically in engineering design. As regards undergraduate design education, the University of Cambridge in Britain and Stanford, MIT and Houston Universities in the USA offer excellent programs. It is noteworthy that all these universities introduce design courses early, i.e. starting in the student's third or fourth semester. A general introduction to engineering is given in the second semester.

As far as rock engineering is concerned, Hoek (1987), Salamon (1988) and Sauer (1988) suggested that rock mechanics design in mining and tunneling needs to achieve a unity of purpose by merging together theory and practice in its engineering activities. A comprehensive design methodology for rock engineering, as proposed in the previous chapter, will serve as a vehicle for advancing this goal effectively.

After this background discussion, in order to provide guidelines for improved design education, it is necessary to consider three vital links in the educational chain: high school preparation, university undergraduate training, and graduate/ continuing professional education.

Improved design education must start with improved high-school preparation for two reasons: (1) high schools determine whether or not a student is motivated to accept the *dictum* that good education is a process of life-long learning; and (2) good high school education can free universities from fulfilling 'general education' requirements and providing remedial learning programs thus being able to concentrate on their real mission of higher learning and of training students for professional duties, as is the case in Europe and Japan. Accordingly, high school requirements must be reviewed as prerequisites for a meaningful college education.

5.1 HIGH-SCHOOL PREPARATION

The syndrome of 'a nation at risk' is dramatically evident when we consider that 95% of Japanese high school students perform better on their mathematics and science examinations than the *top* 5% of US students. And this is in spite of the fact that the United States spends on education 6.8% of its GNP, by comparison with 6.5% by Japan and 4.6% by Germany. Total expenditure on education in America by the state, local and federal governments amounted to $353 billion in 1990, including primary, secondary and higher education.

There are many reasons for this state of affairs. For example, the typical American student has less than an hour of homework per day but the Japanese student has two and a half hours or more. In addition, while American youngsters are in class for an average of 180 days per year, students in Japan spend 243 days in school. In Germany the figure is 232 days, in South Korea, 220 days and in Thailand, 200 days. The traditional 180-day school year in the US was established about a century ago, when children were needed during the summer months to work on farms. Now children are not allowed to work on farms as liability precludes all but the farmer's own children!

Another disturbing factor is the drop-out rate: 33-37% in 1990. This four-year rate means that over one-third of the students entering ninth grade in 1985 left school before graduating in 1989. Some high-schools have a drop-out rate of nearly 60%!

The third reason is what the *New York Times* described as the characteristic of US youth in the 1990s: the *Indifferent Generation* which 'knows less, cares less, votes less, and is less critical of its leaders and institutions than young people in the past.' *Boston Globe* (November 11, 1990) called it the *I-Deserve-It Generation*; their ideology is self-centered and expedient: 'Whatever I want, I need; whatever I need, I deserve. Whatever I deserve, I must get by whatever means.'

As reported by *Time* magazine on February 5, 1990, a standardized math test was given to 13-year-olds in six countries in 1989. Koreans did best, with

Americans coming *last* behind Spain, Britain, Ireland and Canada. However, the test also included the statement 'I am good at mathematics.' It is ironic that while only 23% of Koreans answered yes, Americans had an impressive 68% in agreement. *Time* magazine concluded that 'American students may not know their math, but they have evidently absorbed the lessons of the newly fashionable self-esteem curriculum wherein high school students are taught to feel good about themselves.'

All this has a detrimental effect on the quality of the American labor force. As pointed out by *National Review* (April 15, 1991), in the USA about 25% of all workers have IQs below 90 versus only about 15% in Japan. For IQs below 80, the comparable figures are about 10% for the USA and 3% for Japan. These differences are related to Japan's advantages on the factory floor.

There are some encouraging signs on the horizon. Congress has recently introduced bill no. S.2114, *The Excellence in Mathematics, Science and Engineering Education Act*, aimed at making American students first in the world in these disciplines in the next decade. During a hearing on the bill, it was reported* that in a recent survey of 13 countries, US high school seniors ranked 9th in physics, 11th in chemistry, and last in biology. Only about 5% of US undergraduate students earn bachelor of science degrees in engineering compared to 20% in Japan and 37% in Germany. In the case of *engineering doctoral degrees* awarded yearly in the US, 51% are earned by foreign students (at Penn State, the figure is 75%).

It is not surprising therefore that the US Secretary of Education called recently for 'radical reforms' in the nation's schools. The United States now spends $5,208 per student each year, higher than in Japan and Germany and second only to Switzerland. Yet, the nation's average SAT score in 1990 was only 903 out of 1600.

If all the above shortcomings were not enough, they pale before the main problem of secondary education in America: weak high school curricula which do not prepare the students adequately for a college education.

5.1.1 *High school curricula*

High school curricula have been a continuous focus of the education reform movement in the United States inspired in part by the 1983 report *A Nation at Risk*

---

*The number of high school graduates from US public and private schools was 2.8 million in 1988-89. The combined state and federal expenditures at public schools were $199 billion in 1990. Enrollment in institutions of higher education was 12.8 million in 1988 in the USA but only 5.0 million in the Soviet Union. By the year 2000, women are expected to be awarded the majority of associate, bachelor's, master's and doctor's degrees, and 40% of professional degrees. Only 15% of college students complete a bachelor's degree four years after high school, while after six years the figure is only 46%.

by the National Commission on Excellence in Education. Its finding: *Secondary school curricula have been homogenized, diluted, and diffused to the point that they no longer have a central purpose. In effect, we have a cafeteria-style curriculum in which the appetizers and desserts can easily be mistaken for the main courses.*

The Commission's enquiry revealed that a full quarter of credits received by 'general track' high school students were for physical and health education, work experience outside the school, remedial math and English, and 'personal service and development' courses. This part of the curriculum has expanded in many schools at the expense of core academic classes such as American history and algebra.

The National Commission recommended that no American student should graduate from high school without first completing at least four years of English and three years each of social studies, mathematics and science. These are 'noble' standards for this country but in Europe and in Japan, mathematics, science and history are each required for four years in high schools. Yet, since 1983, only three states have attained this lesser US goal: Florida, Louisiana and Pennsylvania. Worse still, the average requirement for American high schools now is only 1.8 years of science. As a result, only 28% of graduating high school seniors have completed the basic courses in English, math, science and social studies recommended in *A Nation at Risk*. A recent survey of 24,000 high schools found that more than 7,000 offer no physics at all, over 4,000 have no chemistry classes, and some 2,000 do not teach biology.

The time the student spends on any subject is no guarantee that the subject will be mastered; the content and quality of high school classes is even more important than their duration. In a more recent report by the US Department of Education, *James Madison High School, a curriculum for American Students* (Bennett, 1987), educational expectations are added to the graduation standards established in *A Nation at Risk*. It describes what four years of English and three years each of social studies, mathematics and science should comprise. It also adds two years each of a foreign language and physical education, and a half-year each of art and music, also suggesting a suitable content for them. This document is intended as a goal, not a statement of federal policy, because the Department of Education is specifically prohibited by statute from exercising direction or control of any school system. From this report, Table 5.1 depicts a 4-year plan for an ideal high school program (36 semester units in grades 9-12 plus optional electives). Each course is two semesters long, except as indicated.

There are more than 15,000 public school districts in the United States. Roughly half of them organize secondary education around a six-period school day which permits 48 semester-units of course work over four years. The other half follow a seven-period day which permits 56 semester units and is clearly preferable academically. High schools in Europe and Japan observe eight academic classes per day, sometimes even including Saturdays! Such school programs

Table 5.1. An 'ideal' high school curriculum for USA. Source: Bennett (1987).

SUBJECT	1st YEAR	2nd YEAR	3rd YEAR	4th YEAR
ENGLISH	Introduction to Literature	American Literature	British Literature	Introduction to World Literature
SOCIAL STUDIES	Western Civilization	American History	Principles of American Democracy *(1 sem.)* and American Democracy & the World *(1 sem.)*	
MATHEMATICS	Three Years Required From Among the Following Courses: Algebra I, Plane & Solid Geometry, Algebra II & Trigonometry, Statistics & Probability *(1 sem.)*, Pre-Calculus *(1 sem.)*, and Calculus AB or BC			
SCIENCE	Three Years Required From Among the Following Courses: Astronomy/Geology, Biology, Chemistry, and Physics or Principles of Technology			
FOREIGN LANGUAGE	Two Years Required in a Single Language From Among Offerings Determined by Local Jurisdictions			
PHYSICAL EDUCATION/ HEALTH	Physical Education/ Health 9	Physical Education/ Health 10	*ELECTIVES*	
FINE ARTS	Art History *(1 sem.)* Music History *(1 sem.)*			

are also in existence in the United States but they are mainly private schools. In this case, to graduate a student must complete four years of English; four years of social studies featuring three years of history and one year of geography; three years each of math, science and a foreign language; one year of computer science, a quarter year of community service; and three additional college level credits. Electives in art, music, physical education and health complete this challenging course selection.

For comparison, Table 5.2 shows a European high school schedule while Table 5.3 depicts one followed by a superior student at the State College High School (there are up to 35 class periods in a week with a minimum of 26 required; the seven-period day does, however, allow for additional extra-mural activities). This is the kind of curriculum that could well serve the future design engineer!

Table 5.2. Grybow High School in Poland. Four-year curriculum.

Period	Subject	1st year (Gr. 9)	2nd year (Gr. 10)	3rd year (Gr. 11)	4th year (Gr. 12)
1	Literature	Introduction to Polish Literature	Polish Poetry and Prose	English, French and German Literature	Russian Literature
2	Mathematics	Math 1: Algebra	Math II: Trigonometry	Math III: Calculus	Math IV: Par. Diff. Eqns
3	Physics	Physics I	Physics II	Physics III	Physics IV
4	Biology	Botany	Zoology	Human Biology	Darwinism
5	Science	Chemistry I	Chemistry II	Chemistry III	Astronomy
6	History	History I: History of Poland	History II: European History	History III: World History	History IV: Contemporary History
7	Social studies	Geography I	Geography II	Geography III	Geology
8	Languages	Russian I	Russian II	Latin I	Latin II
9	Other	Religion	Religion	Cadets	Cadets
Extra-mural (Saturdays)		Phys. Ed.	Phys. Ed.	Phys. Ed.	Phys. Ed.

Table 5.3. State College High School. Four-year curriculum for a superior student (1985-1989). (Class rank 11/524).

Period	Subject	1st year (Gr. 9)	2nd year (Gr. 10)	3rd year (Gr. 11)	4th year (Gr. 12)
1	English	Adv. English: Intro. to Literature	English 10: Intro. to World Literature	Adv. English: American Literature	AP English: British Literature
2	Social Studies	Adv. Social Studies: History and Geography of USA	Adv. World Cultures: European History since 1500	AP American History: Government & Citizenship	Adv. Reading Adv. Writing in Contemporary Society
3	Mathematics	Adv. Geometry*	Adv. Trigonometry and Analysis	Adv. Calculus (B–C)	Adv. Computer Science
4	Science	Biology	Adv. Chemistry	Physics	Physics II
5	Foreign Language	Spanish II	Spanish III	Spanish IV	
6	Physical Ed. and Health	Phys. Ed. Health Ed.	Physical Education	Phys. Ed. Driver Ed. Health Ed.	Physical Education
7	Other	Latin I	Latin II	Drafting and Engineering Graphics	College Math (PSU) 230H Calc/Vect. 251H Par. Diff. Eqns.
Extra-mural		Clarinet and Band	Junior Soccer Team	Varsity Soccer Team	Varsity Soccer Team

*Prerequisite Algebra II taken in Grade 8.

What can be done to improve the state of public education in America? Dr. Derek Bok, President of Harvard University has these suggestions (*Harvard Gazette*, September 7, 1990):

1. Make school systems less centralized and place more control over the curriculum, hiring and firing of teachers, in the hands of local schools;

2. Adjust teacher salaries to offer monetary incentives for excellence in teaching;

3. Offer better training for principals (school directors) and emphasize the role of principals as educational leaders;

4. Improve the quality of teachers through more effective training and recruiting practices;

5. Adopt more rigorous high school curricula.

High school education is but the first step in life-long education and its importance should be seen in this perspective. In fact, long ago, Plato outlined the need for and benefits of life-long education, saying that true education comes from an introduction to technical subjects – particularly mathematics – when young, followed by a period of practical experience in the world, followed in turn by the study of philosophy. Wrote University of Texas Chancellor Hans Mark (1988): 'Such a blend of (engineering) training, (industrial) experience and study (of history, philosophy, and literature) would enable people in their 50s to approach genuine wisdom.'

5.2 UNIVERSITY EDUCATION

The next weak link in the educational chain of a future design engineer is the undergraduate education at an engineering university. America has some of the finest universities and colleges in the world (see Table 5.4) and its engineering

Table 5.4. Rating of major universities in the USA (*1991 College Guide*, US News and World Report, June 3, 1991). Survey included 1,374 universities and colleges; ratings were based on academic reputation, student selectivity, faculty resources, financial resources and student satisfaction.

Private universities

1. Harvard	7. Duke
2. Stanford	8. Dartmouth
3. Yale	9. Cornell
4. Princeton	10. Columbia
5. California Inst. of Techn.	11. Univ. of Chicago/Brown U.
6. Massachusetts Inst. of Techn.	12. Univ. of Pennsylvania

Public universities

1. California, Berkeley	8. Washington
2. California, Los Angeles	9. California, San Diego
3. Virginia	10. Pennsylvania State Univ.
4. North Carolina	11. Virginia Polytechnic Inst.
5. Michigan	12. Indiana
6. Illinois	13. Minnesota
7. Wisconsin	14. Texas

Table 5.5. Rating of engineering universities in the USA (*US News and World Report*, April 29, 1991). Survey included 193 accredited engineering universities; ratings were based on academic reputation, practicing engineers' reputation, selectivity, research activity, and professorial scholarship.

1. Massachusetts Inst. of Techn.	11. Georgia Inst. of Techn.
2. Stanford	12. Pennsylvania State Univ.
3. Illinois	13. Ohio State
4. California Inst. of Techn.	14. Wisconsin
5. Michigan	15. Rensselaer Politechnic Inst.
6. California, Berkeley	16. Univ. of Southern Calif.
7. Purdue	17. Princeton
8. Texas, Austin	18. Texas A & M
9. Cornell	19. California, Los Angeles
10. Carnegie Mellon	20. Virginia Politechnic Inst.

Rating of engineering programs in the USA having rock engineering components

Civil engineering	Mining engineering
1. California, Berkeley	1. Colorado School of Mines
2. Illinois	2. Virginia Politechnic Inst.
3. MIT	3. Pennsylvania State Univ.
4. Stanford	4. Missouri
5. Texas, Austin	5. Kentucky

Petroleum engineering	Geological engineering
1. Texas, Austin	1. Colorado School of Mines
2. Texas A & M	2. Missouri
3. Stanford	3. Minnesota
4. Louisiana	4. Arizona
5. Oklahoma	5. Utah

universities number 269 of which 204 have been regularly evaluated and rated (see Table 5.5 from *US News & World Report*, October 15, 1990).

In 1989, engineering degrees granted by US universities were as follows:

B.S. degrees: 68,824 (foreign students: 5,719)

M.S. degrees: 26,412 (foreign students: 7,474)

Ph.D. degrees: 5,017 (foreign students: 2,539).

These are high numbers by comparison with other nations (only Japan and the Soviet Union graduate more engineers than the USA) and there are plenty of superb universities and colleges in this country where students get an excellent education. But, on the whole, our colleges are educationally undemanding and economically wasteful (*Washington Post*, October 7, 1990: 'Higher Education's Low Standards' by Robert J. Samuelson). Consider these facts:

1. The attrition rate among college students is enormous. Out of entering freshmen at four-year colleges, only 46% will have earned a bachelor's degree after *six* years.

2. Two-thirds of college professors say their university increasingly teaches what students should have learned in high school. University administrators see themselves as the victims of poor high schools.

3. The value of degrees is suspect: nearly 30% of bachelor's degrees are in

'business' or 'communications,' double the rate of 20 years ago. In general, these degrees do not make significant intellectual demands on students or provide important technical skills (such as engineering).

4. Most colleges are obsessed with surviving. They subtly lower academic standards to ensure the flow of students and subsidies.

5. The badge of successfully completing high school is the ability to go to college. There are 3,600 colleges and universities in the USA (2,100 are four-year institutions) but only some 200 are truly selective in the sense that they reject applicants. About 60% of high school graduates go on to some college.

6. Higher education now accounts for 40% of all US educational spending.

7. Of last year's 12,247,000 students, over 51% were female, 43% were 25 years old or older, and 45% attended part time.

5.2.1 *Undergraduate engineering curricula*

To determine the status of engineering education at our universities, we should first answer a question: 'What is good education'?

At Harvard University, good education starts with a prescribed *core curriculum*. This curriculum program differs considerably from programs of 'general education' practiced by other universities. The core curriculum at Harvard does not define intellectual breadth as mastery of a set of *Great Books* or the digesting of a specific amount of information, or surveying current knowledge in certain fields. Rather, the Harvard core curriculum seeks to introduce students to the major approaches to knowledge in areas that the professors consider indispensable to undergraduate education. It aims to show what kinds of knowledge and what form of inquiry exist in these areas, how different means of analysis are acquired and used, and what is their value.

Each student is required to fulfill the requirements in eight of the ten subdivisions of the core. The areas are: foreign cultures, historical study, literature, arts, moral reasoning, science, social analysis, and quantitative reasoning. Typically, for a non-science major, the science core areas (two segments) are compulsory while literature and arts would be mandatory for science and engineering majors. An example of the core subjects is given in Table 5.6.

Table 5.6. An example of core curriculum at Harvard.

Foreign cultures A4:	Slavic romanticism
Historical study A11:	Development and understanding – the historical origins of the inequality of nations
Historical study B35:	The French Revolution – causes, processes, and consequences
Literature and arts A32:	Dialogues of friendship from Socrates to Shakespeare
Literature and arts B17:	The visual arts – theoretical and practical explorations of the studio arts
Literature and arts C55:	Opera: Perspectives on music and drama
Moral reasoning 32:	Reason and evaluation: Aristotle to Kant
Science B27:	Human origins and evolution

In essence, undergraduates must devote a quarter of their studies to the courses in the core curriculum (eight one-semester courses). To graduate from Harvard in any 'concentration' (i.e. a major) requires a total of 32 one-semester courses (equivalent to 116 credits at other universities) of which 16 courses are in the student's major. This leaves another eight one-semester courses as electives which may, but do not have to be, in the area of 'concentration.' However, to satisfy accreditation in engineering, usually four elective one-semester courses in engineering are needed by students in this major.

At the Massachusetts Institute of Technology (MIT), one of the premier engineering schools in the world, good undergraduate education in engineering is defined as one which *provides graduates with the attitudes, habits and approaches to learning that would ensure a lifetime of technical competence, social contribution and personal fulfillment.*

The goals of engineering education for MIT graduates are:

1. Have a firm foundation in the sciences basic to their technical field and a working knowledge of current technology in a field of specialization;

2. Have an opportunity to exercise ingenuity and inventiveness on a research project and perform engineering synthesis on a design project;

3. Understand the economic, cultural, political, social, and environmental issues surrounding technical development;

4. Begin to understand the diverse nature and history of human societies, as well as their literary, philosophical and artistic traditions;

5. Attain good oral and written communication skills;

6. Acquire the motivation for continued self-education for career-long learning.

It is clear from the above that undergraduate engineering education should be broadly conceived. In contrast, graduate education at a master's program in engineering should allow students to learn in-depth the technology of their specialty and the elements of professional practice including emphasis on design, development, production, and economics. A master's degree should not necessarily be viewed as an intermediate step towards a doctorate.

A document entitled *A National Action Agenda for Engineering Education* was published by the American Society for Engineering Education (ASEE, 1987). It lists these recommendations among others:

1. The overburdened curriculum. The four-year undergraduate engineering program should be designed by engineering professors to provide a knowledge base and capability for career-long learning. It should include the appropriate sciences and mathematics and the fundamental concepts of analysis and design. Repackage the overburdened curriculum.

2. Practice-oriented graduate programs. Advanced degree programs focused on engineering practice should be vigorously developed by engineering professors in a variety of technological specialties to complement the currently available research oriented advanced degree programs in engineering.

3. Design/Manufacturing/Construction. The principles of engineering design, leading to the manufacturing and construction process, should be given a central role in undergraduate curricula. It is essential that professors organize courses in the methodology of engineering design.

4. Disciplinary specialization. Delay much of the disciplinary specialization in the undergraduate program until the graduate years and concentrate on fundamental engineering principles and practice.

These recommendations imply that a B.S. degree is no longer the entry-level degree for engineers in America and that it should be supplemented by an M.S. degree. This, in fact, would bring us towards the European standard of a five-year engineering degree (*Diplom-Ingenieur* in Germany) which includes a prerequisite of practical experience.

In a way, this may also be interpreted as the first step towards a professional degree in engineering similar to what exists in medicine and law. In fact, attempts in this respect go back to 1817 when Sylvanus Thayer introduced a formal engineering curriculum at the West Point Military Academy. He endowed a graduate school of engineering at Dartmouth College in his later years intending that students would enroll in the Thayer School's two-year program after a three year bachelor of arts degree. Today, the Dartmouth 3-2 program is unique in the country; other universities experimented with this model but reverted, largely under economic pressures, to four-year curricula.

However, while European engineering graduates complete a five-year program consisting entirely of science and engineering subjects (see Table 5.7), the American engineering universities, including MIT, Stanford and Penn State, not only use a four-year program but also require that engineering students take 20% or even more of their courses in arts, humanities and social sciences. This means that out of a total of 128 semester-credits typically needed for graduation in the United States less than one-third are in the engineering major (typically 31 credits).

In some countries, renewed emphasis is being placed on design education. In Britain, a special journal *Engineering Design Education* is published semi-annually by the Design Council (since 1985). In Israel, engineering curricula are being changed (Technion, 1987) to accommodate greater emphasis on design and, ironically, to implement US recommendations by ASEE (1987). As shown in Table 5.7, Israel's only engineering university is adopting a reduced four-year curriculum of 144 semester credits (down from 165 previously) which devotes 30-35% of credits to mathematics and natural sciences, 35-40% to engineering sciences, 15-20% to design and computer technology, and 10% to humanities, social sciences, communication and the English language. The changes in the curriculum involved adopting these policy concepts:

1. Postponing extensive disciplinary specialization to the graduate level.

2. Reducing the portion of elective subjects, both free electives and departmental electives, and replacing them with coherent study programs clustered around major themes.

Table 5.7. Comparison of engineering curricula at universities.

Institution	USA		Israel	Poland*	Germany**	So. Afr.
	Penn State	UC Berkeley	Technion	Gdansk TU	TU Berlin	Johannesburg
Source****	1	2	3	4	5	6
Total credits	128	122	144	153	172	182
Total in major	31	43	48	74	64	86
Length of study	4 yrs	4 yrs	4 yrs	5 yrs	5 yrs	4 yrs
Mathematics	16	16		21		
			49		48	69
Basic sciences	22	23		30		
Engineering sciences	35(17)***	41	54	56	60	68
Engineering design	19(14)***	21	27	30	38	35
Foreign language	0	0	3	6	0	0
Arts/Humanities	18	18		0	0	0
English/Writing	6	0	11	0	0	0
Communications	6	3		0	0	6
Other (Phys. Ed.)	6	0	0	10	10 (law)	4

*In Poland, 16 weeks of practical experience is required for graduation: 6 weeks after year 2 and 3, and 4 weeks before the 5th year.

**In Germany, practical experience is required as follows: 8 weeks of experience 'prior to entrollment'; 5 weeks before interim exam; 13 weeks prior to final exam. In mining, 200 shifts of experience are required, incl. 110 prior to enrollment.

***Credits taken in the engineering specialization (i.e. major)

****Sources: 1. PSU Bulletin (1990/91), p.135. 2. *Berkeley Engineering*, 83, 1989, p.34. 3. See reference: Technion (1987). 4. Informator Politechniki Gdanskiej (Gdansk Technical University Bulletin) and personal study, 1988. 5. Böde, K. Bergbaustudium in der Vereinigten Staaten und in der BR Deutschland. *Erzmetall* 43, p.502 (1990). 6. University of Witwatersrand, Johannesburg, *A Report on the Undergraduate Programme in Mining Engineering for the Advisory Committee*. Chamber of Mines of South Africa, Johannesburg, April 18, 1990, p.15.

3. Introducing honors programs for above average and highly motivated students.

4. Revising teaching methods and learning habits with emphasis on individual study.

5. Encouraging the top third of the student body to continue studies immediately for the master's degree.

Moreover, one significant recommendation for the future is to move science and mathematics material from university to high schools, because high school education is being upgraded continuously.

Technion (1987) places special emphasis on engineering design. It considers that the design element of the curriculum urgently needs to be strengthened by adding courses whenever possible on the fundamentals of design, and by incorporating open-ended problems in most engineering science subjects. At the same time, the curriculum should be purged of archaic empirical design approaches.

5.2.2 *Engineering design education*

In essence, the current overall state of education in the United States is characterized by (1) an ailing high school education system, (2) falling standards in undergraduate college education, and (3) a superb graduate university education system. How does this 'bad news' and 'good news' mixture affect engineering design education? Certainly, poor high school preparation of students in spite of massive expenditures (about $130,000 per classroom*) has a major detrimental effect on undergraduate college education, particularly at the engineering universities, which compound the matter by limiting the number of engineering subjects in the curriculum to about half of what is required in Europe and by not requiring practical experience before graduation as is the norm in other countries. American graduate education is excellent and is the envy of the world which is the reason why masses of foreign students flock to US universities. However, fewer and fewer American citizens elect to pursue graduate degrees in engineering and this does not improve the state of engineering design education in the United States.

With the above background in mind, what guidelines might be given for engineering design education? First, consider these problems: the most recent ABET general criteria specify (ABET, 1987) a minimum of one year of engineering sciences but still only a half-year of engineering design (i.e. 16 semester credit hours in a program of 128 credit hours). Is it enough?

Moreover, most ways of teaching design concentrate on the technical aspects of the problem and avoid dealing with the broader issues (social, political, economic) and the implications of the design process. In general, far too much emphasis is placed on the object being designed rather than on the strategies of the design process itself. Another major problem is that few universities are able or willing to begin design education in the freshman year. In most cases, design classes are offered at junior level.

'Although design encapsulates the essence of the engineering profession and speaks to its ultimate objective, the mere mention of the term in academia brings forth an array of descriptions and interpretations,' wrote Evans et al (1990) after a special study of this issue. The study concluded that despite three decades of learned reports that recommend more design throughout the engineering curriculum, including even a half-year design course, the topic remains controversial. Clearly, a vision for engineering design education is missing.

An indication of the desperate state of design education is the fact that the Boeing Company of Seattle, the world's major aircraft producer, decided to take

---

*It is still a mystery why this $130,000 per classroom (i.e. 25 students a $5,208 each) brings such mediocre learning results. After all, if a teacher's salary is $30,000 what happens to the remaining $100,000? Is a major share of this vast sum wasted on excessive school administration and bureaucracy? Apparently, only 20% of the per classroom amount is actually spent on the class!

matters into their own hands by instituting several in-house programs aimed at improving aircraft design education and by conducting courses and training during 1989 and 1990 for over 150 top students from various major universities. As a result of these activities (McMasters and Ford, 1990) they have learned that:

1. Almost none of the students (seniors in aerospace engineering) seemed to have a realistic idea of how a company like Boeing goes about designing a new airplane, let alone how one is manufactured.

2. None of the students had been prepared by their education to deal with the open-ended design-related problems posed by Boeing, nor did they seem to have much experience of working in groups to solve problems collectively.

3. Many of the professors teaching design, who were contacted by Boeing, admitted to having very limited, if any, design experience.

4. In teaching design, most professors concentrate on the technical aspects of the problem and avoid dealing with the broader issues and implications of the design process. Far too much emphasis is placed on the objects being designed rather than on the strategies of the design process itself.

5. When it comes to the very definition of design, even *design* faculty, those who are often segregated from *analysis* faculty by the courses they teach, have trouble articulating the elusive term *design*.

6. It is essential to develop a modern approach to teaching design as a fully integrated part of the curriculum (including the core engineering courses) from the freshman year on. Design cannot be taught in one course.

This last observation is particularly important and has already been put into practice in an innovative way by the University of Rhode Island (Viets, 1990) where freshmen engineering students are assigned a design project that they work on in various courses until they graduate four years later. At the University of Michigan, engineering students are faced with actual open-ended design problems in each required course in their major (Ernst and Lohmann, 1990).At the Arizona State University, design education starts with freshmen (McNeill et al. 1990) because 'freshmen students must be respected as an important resource to be nourished.' The same philosophy is used by the Colorado School of Mines (Olds et al. 1990). Moreover, as mentioned earlier, Crochetiere (1987) reported that Tufts University has had a Department of Engineering Design for over 20 years (!) and currently offers M.S. and Ph.D. degress specifically in engineering design.

Dixon (1991) pointed out, in two thought-provoking papers, that the *intellectual and curricular stagnation in design education* is due to the fact that the design education community never developed a consensus on what constitutes the *fundamentals* of the field of engineering design (in stark contrast to what has occurred in engineering science). Fields that lack a foundation of principles cannot advance intellectually and must rely on collections of specific *ad hoc* heuristics. This situation is due to six basic intellectual mistakes made by the engineering community regarding design education (Dixon, 1991):

Mistake 1. Very little research was done to develop a scientific understanding of the engineering design processes.

Mistake 2. Engineering design is not an art or a skill which can be learned only by experience, but rather is a cognitive, or intellectual, process based on design science fundamentals.

Mistake 3. We do not distinguish clearly betwen analysis and design and, when in doubt, we assume that if students can do analysis, they can do design.

Mistake 4. We often confuse experience with learning, or provide experience without education; not all experience is necessarily good, and it is possible to learn the wrong thing from experience.

Mistake 5. By limiting design education to technical design issues, we omit the rest of the engineering design process. Engineering design involves much more than technical design and the functional issues which can be analysed by engineering science.

Mistake 6. By trying to make ready-to-work designers of students in four years, we avoid deciding what limited part of engineering design education realistically belongs in the university (i.e. the fundamentals) and what must be left for industry to do.

In addition, an important aspect of design must not be overlooked and this is the element of creativity which plays a crucial educational function (Chaplin, 1989). As discussed in Chapter 2, design is a vehicle for turning creativity into innovation. This calls for creativity enhancement and the curriculum should be structured to allow time for experimenting with new ideas and developing new skills such as those discussed in Chapter 7. This would enable students to exercise their curiosity and learn how to search for knowledge as background for the generation of new ideas. Alas, current curricula seldom feature such an approach.

To correct these mistakes and develop a new engineering design curriculum, one must identify and teach the *fundamentals* of engineering design (NRC, 1991). Teaching current design practice is a second priority. We can determine the fundamentals to be taught by carefully observing the scientific foundations for the current best practices in the most competitive companies and the relevant results obtained so far from research into engineering design science.

Fundamental knowledge is one that is sufficiently general to provide at least partial intellectual support for a number of specific applications and for the future learning of new applications and other fundamentals. Four categories of fundamentals can be identified (Dixon, 1991): (1) knowledge *about* a concept, fact, principle, methodology or technology; (2) knowledge of *how to use* these aspects; (3) behaviors; and (4) attitudes, values, and beliefs.

Table 5.8 lists fundamental topics in all these categories. There are a number of ways to organize the set of fundamental topics into a set of courses for a curriculum. One possible organization (Dixon, 1991) is shown in Table 5.9.

Table 5.8. Proposed fundamentals of engineering design science derived from current best practices and research results (after Dixon, 1991).

1. The business context of engineering design
   Knowledge of essential steps in product realization processes and engineering economics.
   Understanding of relationships of marketing, finance, manufacturing, management, and personnel issues.
   Understanding of quality, cost, and time to market in product realization.
2. Concurrent engineering and team participation principles
   Knowledge of concurrent design process concepts and practices.
   Effective performance on a cross-cultural team, including making decisions.
   Presentation of effective reports.
3. Manufacturing and construction
   Knowledge of manufacturing and construction processes, their physics, economics, and practices.
   Knowledge of design for manufacturability, design for constructibility, and design for life cycle.
   Knowledge of statistical process control methods.
4. Analysis and prototyping
   Knowledge of common computational prototyping methods.
   Knowledge of analytical modeling processes. Ability to develop appropriate idealized models of designs for analytical purposes.
5. Statistics
   Knowledge of and ability to use probability, statistics, decision theory, and design of experiments in principle.
6. Design theory and methodology
   Knowledge of the most prominent descriptive, prescriptive, and computer-based models of design processes.
   Competence to perform design using Pahl and Beitz design methodology.
7. Component design and optimization
   Ability to design, evaluate, and redesign components at conceptual, configurational, and parametric stages relating trial design performance to technical, manufacturing, cost, and other life-cycle issues.
   Appropriate formulation of problems for optimization.
8. Systems
   Knowledge of and ability to use systems design and system integration principles.
   Performance of conceptual, configurational, and parametric design of systems that integrate mechanical, electronic, optical, and computer elements.
9. Computer-aided design
   Knowledge of and ability to use CAD, and computer modeling systems.
10. New information and learning
    Ability to keep informed and learn as needed about new materials, technologies, and processes through reading, discussion, and technical and business conferences.

NRC (1991) published an important report *Improving Engineering Design* which presents a series of recommendations for engineering design education. It states that undergraduate engineering design education must:

1. Show how the fundamental engineering science background is relevant to effective design;

2. Teach students what the design process entails and familiarize them with the basic tools of the process;

3. Demonstrate that design involves not just achieving a function but also leads to productivity; and

Table 5.9. An organization of fundamentals into courses (after Dixon, 1991).

Course title	Subject content of the design fundamental
1. Product realization processes in business organizations	Generic product realization process. Taxonomies (classification) of design, product development, and decision problem types. Introduction to marketing, management, finance, and personal issues and concepts.
2. Manufacturing and construction processes	Basic physics and practical factors in the most prominent manufacturing and construction processes; design for manufacture and constructibility and cost estimating.
3. Design of components	Parametric design and redesign of components employing optimization, and systematic design methodology. Configuration design of components; cost estimation.
4. Concurrent multi-functional team design	Individual and group issues and principles for effective participation in multi-functional design teams.
5. Case studies of best design practice	Case histories using the case method approach.
6. Information retrieval and learning	How to find information about materials, processes, new methodologies and technologies using libraries, computer searches, telephone enquiries, vendors, colleagues, consultants and commercial registers. How to learn on one's own to employ new methodologies and technologies using textbooks, journals, short courses and technical meetings.

4. Convey the importance of other subjects such as mathematics, economics, and manufacturing.

Graduate design education should be directed toward:

1. Developing competence in advanced theory and methodology.

2. Familiarizing graduate students with state-of-the-art ideas in design, both from academic research and from worldwide industrial experience and industrial research.

3. Immersing students in the entire spectrum of design considerations, preferably during industrial internships; and

4. Having students perform research in engineering design.

Owen (1990), writing from the perspective of design education in Britain, suggested that graduate level design programs should be differentiated to offer opportunities for professional mastery in a Master of Design program, and for research in Master of Science and Ph.D. programs. This would provide the incentive for universities to pursue excellence in specialized design areas and achieve depth with quality. This approach has in fact, been used for a long time in the USA by the Graduate School of Design at Harvard University.

NRC (1991) emphasizes that a continual stream of design-oriented doctoral graduates with new design knowledge is needed to supply professors who can teach undergraduate engineering design. In addition, because textbooks that provide a comprehensive insight into the field of engineering design are rare, an urgent effort must be made to rectify this situation.

If the recommendations by NRC (1991) and by Dixon (1991) are followed, better engineering design education will improve the practice of engineering in the United States. In a few years, universities could begin to graduate students whose knowledge of engineering design, contact with industry during their schooling, and awareness of good design practices would attune them to the needs of industry and the realities of engineering design as well as dispose them to continuing education throughout their careers. These graduates could augment and eventually replace a generation of designers who received limited coherent engineering design education. Moreover, if the NRC (1991) recommendations are followed, students who emerge from graduate engineering design programs familiar with current advances in the theoretical foundations of design and forefront methodologies will not only contribute to engineering practice, but also be prepared to create new design tools, teach design to the next generation of students, and conduct research in design.

5.3 HIGHER EDUCATION ISSUES IN ROCK ENGINEERING

The subject of higher education in the engineering programs having rock engineering components has received world-wide attention. In civil engineering, Tarricone (1990) and in mining engineering Shaw (1989) lamented the unsatisfactory current situation in their disciplines: the civil and mining engineering professions are mired in image and recruitment problems, the university curricula are under fire and the best students are turning to business and law as engineering has lost its luster. They pointed out many issues needing resolution including ensuring equivalent educational standards between nations engaged in global competition as well as the amount of practical experience which must go along with the academic training of engineers. Rauhut (1989) presented German aims for the training of young mining engineers which sets a fine example for top quality engineering education. Fundamentals of the training of mining engineers in Germany are practical experience, a broad knowledge of the natural sciences and engineering as well as a basic knowledge of economics and law. In this way, not only do they become professional engineers but they develop social responsibility towards employees, society and the environment.

Tarricone (1990) points out an established significant fact reflecting the *20/80* theory: generally, engineers only apply about 20% of what they learn as undergraduates; the other 80% of knowledge comes from job experience and graduate studies. In view of this situation what should be the university education of design engineers in the field of rock engineering?

Fettweis (1989) and Gergowicz (1990) addressed this very problem while Gentry (1990), in a thought-provoking address, provided a framework for the role of engineering design in the future curricula for rock engineers. Gentry stated: 'We must come to realize that our role as educators is not simply to transfer knowledge, that it is no longer sufficient to prepare students for today's technologies. Students must be prepared to grow and learn in order to meet the problems of tomorrow. We must distinguish between training technicians and educating engineers for positions of future leadership in the industry.'

Gentry (1990) suggested that the quality of the entire educational experience at the undergraduate level is as much a function of the educational process as it is of the content. He recommended that we must challenge not only the content of rock engineering-related curricula but we must also challenge the curricular structure, as well as the education process in general.

In essence, the challenge to education is to develop in students the skills that provide for an immediate, productive career start, while – at the same time – add the scientific base that, together with a lifelong thirst for learning, will stimulate intellectual growth and the ability to lead technological change. Gentry (1990) believes that this can be achieved by ensuring the presence of three specific components in rock engineering education. One is the need for 'internationalizing' the curricula. The graduates must be able to function as productive, responsible international citizens who can recognize that engineering and scientific factors alone no longer serve as the sole basis for investment decisions in America's major mining and construction corporations.

The second component concerns the teaching and structuring of subject areas contained in the lower division of engineering curricula (e.g. mathematics, basic sciences and fundamental engineering sciences). They must be re-thought and made more meaningful to engineering students. Currently, some of these courses often cause students to withdraw from the study of engineering because they experience little if any relevance to the practice of engineering, even after two years of formal study! Entry level courses should focus on thinking, analysis, synthesis, critical reasoning, and less on memorization of facts. The fundamental importance of the interdependent nature of the sciences, mathematics and the practice of engineering must be clearly understood by the students.

The third component, according to Gentry (1990), is engineering design as the essence of engineering education. Engineering design must be seen to represent an *outcome* of the entire undergraduate engineering education process. An emphasis on design, innovation and creativity is to reflect the principle that the quality of the design experience is more important than the quantity requirement often used.

Will there be enough time in a four-year program to accommodate these changes in the rock engineering curriculum of the future? Or is a five-year degree structure inevitable? In fact, the four-year engineering programs now offered in America range between 4.5 and 4.7 years. Engineering students are currently experiencing a five-year educational program whether universities want to admit it or not! Thus, let us restructure the engineering curriculum over five years.

With this situation in mind, how should we structure an ideal rock engineering curriculum which would ensure sound design education? It is clear that such an undertaking should be based on three premises: (1) appropriate high school preparation, as demonstrated in Table 5.3; (2) a framework of meaningful engineering education, as advocated by Gentry (1990); and (3) a design education component, as proposed by Dixon (1991).

An example of an effective but tough rock engineering curriculum is given by Gergowicz (1990). The program is of five years duration, divided into ten semesters (see Table 5.7, item 4). The tenth semester is devoted to an M.S. thesis. The first two years cover the fundamentals of engineering (mathematics and basic sciences as well as an introduction to rock engineering). From the fifth semester on, engineering courses predominate with the subject of rock mechanics being taught over three semesters plus playing a major part in the design courses. Typical to a European technical university, Gergowicz (1990) does not include any subjects in liberal arts and humanities, although foreign languages are required. In true European tradition, 31 hours of class time per week (!) are expected of the students with no choice of scheduling being permitted! However, his curriculum – being so highly oriented towards technical courses – would not satisfy the criteria posed by Gentry (1990) and is rather weak on the component of design.

Well, what then would be the best engineering curriculum for our imaginary design engineer for the 21st century? Table 5.10 contains one possible arrangement of courses to achieve this purpose, and the author invites a vigorous discussion on this topic.

Table 5.10. An engineering curriculum with a strong design component.

		Remarks
Length of studies	5 yrs	Culminates in MS degree
Total credits	160	16 credits per semester
Total one-semester courses	55	One course = ~3 credits, ex. labs
Credits in the major	50*	Engineering specialization
Mathematics	16	Four one-semester courses
Basic sciences	25	Seven one-semester courses and two laboratories
Engineering sciences	40(20)*	Ten one-semester courses and three laboratories
Engineering design	28(12)*	Incl. 16 credits (six courses) of design core from Table 5.9 plus six design courses in the major
Graduate courses	12*	Four courses
Professional report (not a research thesis)	6*	Two semesters
Foreign language	6	
Arts and humanities	18	
Technical writing	3	Total: 11 courses
Computer proficiency	3	
Other: Engineering graphics	3	

*Credits taken in the engineering specialization (i.e. major), 50 credits is equivalent to 20 one-semester courses.

Example 99

5.4 EXAMPLE

The purpose of this example is to show that even before the recommendations listed in Table 5.10 are put into effect, one can obtain a high quality engineering design education in the United States if an effort is made to plan the student's academic schedule. While such renowned engineering schools as MIT or Stanford are known for their engineering design programs, this example involves a university not traditionally thought of as an engineering institution.

Using a 'crystal ball' to see what would be the most desirable profession in the year 2001 and which university to select, we may be guided by Lewis and Kingston (1989)* to select environmental engineering at Harvard University. Environmental engineering is widely recognized as one of the most promising careers of the future as environmental issues will be the governing factors in the mining and construction industries and will require innovative solutions and high design skills on the part of design engineers. Since the author is particularly familiar with Harvard, the example of a degree in environmental engineering, an accredited program at Harvard University, was selected.

Harvard requires 32 courses for graduation, comprising the core curriculum (8 courses), an environmental engineering 'concentration' (i.e. specialization) amounting to 16 courses, plus eight electives: two courses to broaden the student's education and six courses to satisfy the engineering accreditation criteria. A possible curriculum is depicted in Table 5.11. These are actual courses from the 1990/91 Harvard bulletin.

Let's imagine that our student, we will call him Alec B., decided to use his electives to obtain the best engineering design education possible at Harvard. In the process, he will proceed to secure an M.S. degree in design studies during his fifth year at Harvard before joining the corporate world as an environmental design engineer with a particular interest in rock engineering. He will graduate in time to be well prepared to tackle the illusive problem of designing a high level nuclear waste repository due in operation by the year 2010!

Note that Alec the environmental engineering student has received credit for some advanced courses taken at high school in terms of the Princeton Advanced Placement examination which exempted him from the first six courses in Table 5.11. For these courses he substituted those taken at MIT giving him a total of 30 courses: 22 courses from Table 5.11 plus the core curriculum courses (Table 5.6).

A master of science degree at Harvard is of one year duration and requires 8 graduate courses and no thesis. Alec will take this degree at the Graduate School

*This reference shows that 54% of corporate leaders and 42% of government leaders had graduated from 12 universities: Harvard, Stanford, Yale, Chicago, MIT, Cornell, Pennsylvania, Princeton, Johns Hopkins, Columbia, Northwestern and Dartmouth.

Table 5.11. An example of environmental engineering curriculum accredited at Harvard (including elective courses at MIT from intercurriculum exchange).

Mathematics 21a:	Multivariable calculus
Mathematics 21b:	Linear algebra and differential equations
Applied Math. 105:	Partial differential equations
Physics 11:	Mechanics
Physics 12:	Electricity, magnetism and waves
Chemistry 10:	Principles of chemistry
Computer Science 50:	Computer programming
Engineering Sci. 101:	Applied statistical techniques for modeling
Engineering Sci. 102:	Introduction to decision and control
Engineering Sci. 121:	Introduction to solid mechanics
Engineering Sci. 123:	Introduction to fluid mechanics
Engineering Sci. 150:	Electromagnetic communication
Engineering Sci. 154:	Electronic circuits
Engineering Sci. 161:	Robotics and computer vision
Engineering Sci. 162:	Hydrology
Engineering Sci. 165:	Environmental engineering
Engineering Sci. 181:	Thermodynamics
Engineering Sci. 190:	Materials science
Engineering Design 96s:	Design process and team design
Engineering Design 96f:	Individual design project
Example of electives	
General Education 180:	Environmental quality and its management
Government 10:	Introduction to political philosophy
Engineering Sci. 208:	Risk assessment in engineering and biological systems
MIT 18.05:	Introduction to probability and statistics for engineers
MIT 1.32:	Introduction to engineering geology
MIT 1.381:	Rock mechanics
MIT 1.383:	Underground construction
MIT 2.70:	Introduction to design

of Design (GSD) which offers a *Master of Design Studies* program. This program is open to graduates in engineering, architecture, urban planning, computer science, and industrial design. Alec's graduate courses are listed below fulfilling the requirement that six of the eight courses have to be taken at GSD:

1. Eng. Sci. 261: Design of water resources
2. MIT 2.70: Axiomatic design
3. GSD 2104: Introduction to design science
4. GSD 2107: Computer-aided design
5. GSD 2302: Geometric modeling
6. GSD 3400: Individual study-design theory and methodology
7. GSD Design Studio: 1st semester
8. GSD Design Studio: 2nd semester.

CHAPTER 6

Design practice in rock engineering

*Not all experience is necessarily good. We must learn only the best practices –
those used most widely in the most successfully competitive firms.*

John Dixon (1991)

Design practice means the actual design procedures, strategies and tactics used in
the industrial 'real world,' where competition, profit, and sometimes survival are
the order of the day. Some design practices are worthy of admiration and the envy
of competitors, others are just mediocre solutions which have not changed and
improved for years. Accordingly, what the 'practical engineer' does in the
industry is not necessarily the best practice, and it should be subjected to close
scrutiny before it is used as an example for others. This chapter not only looks at
the actual design practices but also suggests possible improvements on the basis
of selected case studies.

6.1 A SURVEY OF CURRENT DESIGN PRACTICE

In the past three years, a study was undertaken by the author to determine the state
of design practices in rock engineering. As pointed out before, such cognitive
studies have been performed in other fields (Stauffer et al. 1987; Adelson, 1989)
but none could be found in civil engineering and only one in mining (Sanders and
Peay, 1988).

First, an initial survey was conducted by Kicker (1990) featuring a questionnaire and follow-up interviews with 14 mineral engineering professors at Penn State.
All these persons had been involved in some engineering design in industry.
Subsequently, the author surveyed the views of 89 professional engineers practicing engineering design in mining and tunneling, in the USA and abroad.

The following observations were derived from the initial survey by Kicker
(1990):

A recurring theme in many of the interviews was that a fundamental knowledge
of the problem at hand is crucial to good design and is essential if creativity and
innovation are to thrive. Of those interviewed, 69% said that creativity is being
limited or stifled among mining engineers. Some attributed this stifling effect to

the mining industry, others said that it was the designers themselves who caused the limitation, and some pointed to economics as limiting creativity.

The lack of creativity and innovation in industry was attributed to the following aspects: (1) economics; (2) legal ramifications; (3) lack of a detailed data base of the current state of knowledge; (4) lack of a strong theoretical background; (5) the mining industry not spending an adequate amount of time and money on design.

With one exception, all those interviewed stated that they were not guided in their design practice by any specific design methodology. Mostly, they relied on their experience but pointed out two problems in rock engineering : too much 'data cranking' with no clear purpose, and too much empirically based design.

The following needs were identified:

1. Time is needed to consider and carefully analyze the problem.

2. It is necessary to incorporate 'high-tech' components in our designs and use the latest technology.

3. It is necessary to look at 'the big picture', i.e. everything that is involved in the problem. To this end, a good design methodology is crucial to integrate all aspects of the problem.

4. It is necessary to simplify the problem down to the lowest common denominator and do a comprehensive design on this segment, instead of viewing the problem as too complex.

5. The difference between the *technical knowledge for design* (i.e. tools such as charts, tables, and equations which one needs for design) and the *philosophical methodology of design* (i.e. the procedures to follow as a way of thinking about design and incorporating the technical aspects) needs emphasizing. It is necessary to perform research on the philosophy of design, showing the importance of creativity and decision-making.

The author's survey of design engineers was much larger in scope and in the choice of designers than the preliminary work of Kicker (1990). It was prompted by the excellent studies of Suh (1990) and Hales (1987). In particular, Hales' *Analysis of the Engineering Design Process in an Industrial Context*, which became available in the United States only in 1990, provided a comprehensive framework for reporting the design practices in our industry. Suh's work provided the impetus to search for any design principles that might underlie the selection of a particular design approach.

The author's findings are summarized at the end of this section in a selection of statements by the leading design experts in rock engineering (out of 89 professionals surveyed). Overall, three important conclusions must be singled out:

1. Not all those surveyed agree that rock engineering design is lacking in creativity and innovation; some believe strongly that this is an unfair view, others point to the immense design potential in this field as yet unfulfilled.

2. It was noted that designers do not like surprises because if a problem is anticipated it can be dealt with effectively. Accordingly, identification of potential problems as well as contingency scenarios should be the concern of every rock

engineer. Said one designer: 'If you understand the geologic setting and have considered the design problem from all angles, there will be few surprises.'

3. A staggering finding: not infrequently, design decisions in rock engineering are relegated to the least experienced or even least educated engineers. The experienced supervisors, who usually hold a graduate degree, have little time for design since their main activity is management!

The above observations should be compared with those identifying the current best design practices used at the most competitive manufacturing and construction firms (Dixon, 1991), listed in Table 6.1. Clearly, a progressive company can take a number of effective steps to improve their own engineering design capabilities and thus their competitiveness. They are (NRC, 1991):

1. Recognize the benefits afforded by engineering design and take advantage of them;

2. Implement a comprehensive and coherent engineering design process;

3. Utilize a carefully chosen set of best design practices;

4. Create a supportive environment for design; and

5. Support research and development activities in engineering design.

Are those concepts used in rock engineering? Let us consider a range of views reflecting this field (derived from the author's survey of 89 professionals), followed by four representative design case studies.

6.1.1 *Comments by leading design engineers in mining and tunneling*

Expert 1. The first step in design is to determine the client's needs. Is the mine new, expanding, rehabilitating, closing? Is stability very important, or can local

Table 6.1. Current best design practices, developed by Xerox, Polaroid, Ford, Hewlett-Packard, Carrier, GE and others (after Dixon, 1991).

1. An explicit mechanism for obtaining and considering new and improved product and process ideas from customers, employees (especially employee engineers familiar with new technologies), and marketing information.
2. An explicit mechanism for selecting new ideas for preliminary study relating design, market potential, manufacturing, construction, cost, and company strategy.
3. Explicit use of concurrent design using cross-functional teams to accomplish integration of product function, manufacturing processes, marketing issues, and other life-cycle concerns throughout the product realization process.
4. Explicit well-defined decision points, decision criteria, and decision participants in the product realization process.
5. Constant search for replacement materials.
6. Explicit company-wide concern for quality, cost, and time to market, with special emphasis on quality.
7. Increasing emphasis on systems integration of various technologies.
8. As much strategic integration as possible of computer technologies, including CAD.
9. As much as possible of activity-based costing concepts.
10. Establishment and continual refinement of the design and construction process performance.

instability be treated as a maintenance problem? What is to be mined? Once we define what the client wants, we try and examine what he really needs! The next step, and a very important one indeed, is to gather all the geological information available and drill all the holes you can afford, and map all the exposures one can get at. By studying the geologic setting and developing a working model of the geologic history of a site, the rock mass can be understood. If the rock mass is understood, the design activities will have a reasonable basis, that is, predictable. A predictable design base means no surprises.

We are turning increasingly towards modeling as an important tool in design, however, not to be relied upon solely. Modeling can reveal design flaws and unexpected consequences of a design but is only one tool in the design. Linked with empirical methods, the combination is powerful.

Expert 2. The framework of our design methodology involves incorporating three overlapping and interrelated areas in the design: comparative experience, in situ observation and measurements, and numerical modeling. I regard structural mine design as an evolutionary process that relies mostly on practical experience. Theoretical principles from elasticity and plasticity, as well as concepts derived from soil and structural mechanics, provide important guidelines.

In the structural design we try to visualize the most likely failure mode of the opening (roof, ribs, pillars, floor) and always evaluate any failure, if present, which may shed some light on the design. We also *try* to define the deformations and stress levels that may lead to failure before selecting the working levels at which these must be kept. This involves an assessment of rock mass quality and strength. Back analysis and follow-up during mining are a very important part of the whole process. Safety, economics and ecology are also integral parts of the design.

I believe that mining engineering students lack exposure to design processes in their education. I recommend the use of more guest speakers from industry and government. More of an interdisciplinary approach in their education would be useful in providing a more global view of the design process.

Expert 3. I feel that there is little basis for the statement that *mining is lagging behind other engineering fields by not introducing enough new ideas and design concepts.* The truth is that, in mining practice, both formal and informal design does take place to a considerable degree. Some examples are: (1) equipment selection, (2) equipment scheduling, (3) mine infrastructure, (4) open pit design, (5) planning and design in strip mining.

Concerning *innovation* in design, one basic criterion for applying new technology is to reduce the cost of operations. Simply because some new technology is available does not mean that it should be applied. For mines already in operation it is often very difficult and/or expensive to retrofit them to 'new technology.' I believe that mining engineers have been innovative to a considerable degree given the rather uncertain environment in which they work. As the degree of environmental certainty increases, so does the degree of innovation at the design level.

However, when it comes to the teaching mine design to engineering students, this is generally poorly done in the United States. There are two major reasons for this: (1) very poor teaching materials, and (2) very few professors have ever done a design or been involved in design work. After all, professors are required to have a Ph.D. degree which is a research degree. They specialize in areas other than design. As a result, mine design classes often degenerate into a very descriptive treatment of the subject (e.g. advantages and disadvantages of a given system).

Expert 4. In the mining industry, design must be considered on two distinct levels which, for convenience, I will call macro-design and micro-design. *Macro*-design involves the process of regional geological evaluation, orebody definition, choice of mining methods, decisions on methods of access and overall cost-benefit studies. This design process takes place before the mine is brought into operation and it is the most important design ever carried out on that particular mine because it is done only once. If mistakes are made in the macro-design they are extremely difficult if not impossible to remedy afterwards. *Micro*-design, on the other hand, is a much more obvious process and it can be equated with the normal engineering design processes such as apply to machine components. Once the basic decisions on mine layout have been made during the macro-design stage, all that remains for the site engineering staff is the detailed design of particular excavation shapes, pillar dimensions, rock bolt spacing, etc.

Most of the criticism of the mining industry tends to be directed at the micro-design level since this is readily comprehended and a visible daily activity. There are certainly deficiencies in the micro-design process as applied to mining, however, the main deficiencies are to be found in the macro-design process since there are very poorly defined procedures for choosing the right mining method, for locating the shaft in the right place or for evolving a sound waste disposal policy. Most of the macro-design processes are very heavily dependent upon the experience of a few mining men. Design trends are much more dependent upon tradition and precedent than upon any rational formal design process. Basically, there are very few rational formal design processes available for the macro-design and this is why we continually revert to experience, rules of thumb, and folklore.

Expert 5. In any design, prime consideration is given to past experience in the mine or neighboring mines. Factors influencing the design considerations are discussed. Innovations in design are incorporated after thorough evaluation of the financial aspects. Before any changes are made, it is customary to win the confidence of the mine management, mining unions and the government inspectors. Innovation in mining cannot take place without appropriate assessment of the risks present. Personal experience of the people involved is an essential part of the design package.

Expert 6. Mining can certainly benefit from an infusion of creativity and innovation in the design process! My mining-related experience has been primarily with the surface mining aspects and I can assure you that, despite the recent advances in some areas (i.e. open pit mining, mine tailings disposal, and mine infrastructure), there is still a great need for innovation in the design of these components.

Expert 7. Design is but one stage leading to the successful completion of a tunneling project. Successful designs must integrate the results of engineering investigations for the construction engineer. The design activity must balance the work of the modelers and the analysts to anticipate the response of the underground to the intrusion of an opening and to specify the required ground support system correctly.

Expert 8. Design is the essence and delight of rock engineering. But, over the past twenty years we have collected enough information for the meaningful design of rock structures but we still don't do a good job of designing. Why? Because nobody is researching the design processes in rock engineering; there are no funds for such research and little time either: people are too busy with data cranking! However, the main problem is with design education; our mining and civil engineers are poorly trained in design. European engineers have a much better background in design.

Further comments by design engineers are listed in Table 6.2 giving the excerpts from a 'round table' workshop session conducted by the author on design methodology in rock engineering.

Table 6.2. 'Round table' discussion on current design practice in rock engineering: Comments from design engineers on design methodology.

1. 'The most important thing to remember is that design in mining needs to be practical!'
2. 'I design intuitively!'
3. 'I use a design methodology that I was taught in my first year of college in a mechanical drawing course.'
4. 'We design by consensus of a group since most of our design projects have several people working on them. Generally, we don't use a formal design process.'
5. 'Primarily I design by modifying what has been done in the past. This tends to be easier (less risk), quicker and cheaper.'
6. 'It has been my experience that most new designs are needed when the current system fails. The opportunity to really try something innovative is almost never possible in mining.'
7. 'Previous experience is the main ingredient in my designs.'
8. 'I believe in innovation; I use my experience if appropriate but mostly I innovate when designing.'
9. 'I prefer an unstructured design process.'
10. 'Of course I try a 'structured approach' to design; literature search, going to conferences, and discussions with experts – preferably, friends. However, in the last analysis, it seems that a good solution is more like a lucky coincidence.'

6.2 DESIGN CASE STUDY 1: ROCK TUNNELING

Actual case histories of industrial projects serve the useful purpose of demonstrating what is being done in practice when it comes to executing an engineering venture. Some case studies are examples of good practice, others are examples of what should not be done. So powerful are case studies as a learning aid, that some universities-such as Harvard-have developed teaching programs based entirely on selected cases, mainly in the area of business administration. There is even in existence a World Association for Case Method Research and Application (WACRA), headquartered in Boston, which has held international conferences since 1984 and promotes world-wide usage of the case method (Klein, 1990).

The case method is best defined by contrasting it with the lecture method. Instead of textbooks, the case method uses partial historical descriptions of specific situations enabling people to develop their own unique framework for problem-solving by engaging in critical thinking. Although case studies have been used in one form or another by both law and medicine for a long time, the case method in management and other disciplines is relatively new. The Harvard Business School is well known for its innovative role in this regard.

The case method is increasingly applied in engineering, and agriculture, strategic policy planning, and in higher education, in different parts of the world.

Rock engineering is no exception for the value of case studies to improve the design and construction practices in this field. However, good case histories are scarce in rock engineering and seldom are they well documented. And when it comes specifically to design practices, they are seldom elaborated in sufficient detail even in the more complete case histories.

In the context of this book, a design case study has a three-fold purpose:

1. To demonstrate the state of the art of design practice and how it might have changed over the years;

2. To demonstrate the use of the functional requirements (FRs) and the design (solution) components (DCs) in accordance with the design methodology proposed in Chapter 4; and

3. To point out the benefits of using a systematic engineering design process.

In selecting a design case history for the above purposes, the author wanted to rely on his own experience so that he could comment on what had happened in greater detail than usually found in published case histories. Ten candidate tunneling cases were investigated from the author's files: two railroad tunnels, three highway tunnels, two water-conveyance tunnels and three hydroelectric schemes. These projects were located in Europe, Africa and America and dated from 1972 to present. Some featured very limited design investigations, others involved comprehensive design studies with multi-million dollar budgets.

The project selected for Design Case Study 1 is truly remarkable because it involved very thorough design investigations conducted as long ago as 1975-1978 and the design methodology used then has not really changed or improved significantly over the years, and could be used today as an example of the best practice for any current project. As impressive as this first case study is, it also shows how little progress has been made in design methodology in rock engineering in the past 15 years! The author kept a diary throughout this project recording the technical events, his personal impressions, and his observations of the design team at work.

6.2.1 *Elandsberg pumped storage scheme*

Located not far from the city of Cape Town at the southern tip of Africa, this hydroelectric project was based on the principle of reversible pump-turbines to generate electricity during peak-demand periods while during off-demand hours (nights and weekends) the turbines would be reversed to act as pumps, storing water in an upper reservoir. This 1000 MW project featured an upper reservoir and a dam, an inlet ('headrace') tunnel, penstocks (inclined tunnels conveying water to the turbines) having a water-head of 360 m, an underground power station complex, outlet ('tailrace') tunnels, and a lower reservoir. Excavated rock spans up to 22 m were envisaged which was without precedent in the expected rock conditions.

A general feasibility study was initiated by the owner in 1973 and in March 1974 a report on geological and seismological investigations was issued. It reviewed published data on regional geology and tectonics and on the earthquake history. Based on an assessment of the geological conditions and the seismic activity in the area, the proposed scheme was pronounced feasible. In October 1974, the author's group was requested to undertake rock engineering and design investigations which included planning the site exploration program, rock mechanics tests in the field and in the laboratory, assessment of rock mass quality throughout the site, provision of the design parameters for tunnels and chambers and design recommendations for the spacing, shape, reinforcement and construction sequence of the various excavations. Subsequently, a firm of consulting geologists and one of structural civil engineers were appointed. A contract was also awarded for pilot excavations to start in July 1975. The first underground test site was reached in June 1976. An overview of the rock engineering issues on the project and the scope and purpose of the design studies conducted during the exploration and geotechnical design program were reported by Bieniawski (1976). Figure 6.1 shows the layout of the underground power house complex.

6.2.2 *Geologic site exploration*

A geological cross-section of the scheme is depicted in Figure 6.2. The upper reservoir and most of the penstocks were located in a series of cross-bedded sandstones and quartzitic sandstones with occasional shale and siltstone horizons. The power station and the outlet structures were positioned in rock formations

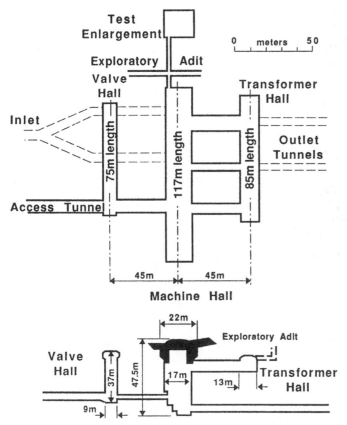

Figure 6.1. Plan and section of the underground chambers.

consisting predominantly of graywacke (90%) with minor amounts of phyllite (5%), quartzite and quartz veins (5%). The phyllites were typically highly sheared, locally grading into slate and generally much weaker than the graywacke.

Faulting was extensive throughout the area, although no major faults extended into the vicinity of the engineering works. The sandstone/graywacke contact was not faulted and it dipped approximately 30° to the east. Three joint sets were recognized from preliminary geological investigations: (1) bedding foliation, near-vertical, striking north-south; (2) vertical cross joints, occurring normal to this strike and showing considerable directional variations; and (3) horizontal, widely spaced non-continuous joints, occurring with increased intensity towards the surface.

In the graywacke strata, the depth of weathering was very high and hard, fresh rock was only encountered between 50-100m below the surface. The depth of weathering was a function of the variation in the water table and the extent of vertical to sub-vertical fracturing. Three structural regions were recognized in the

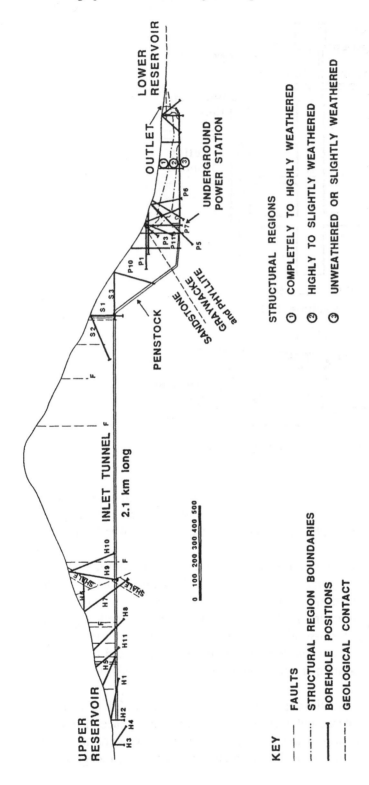

Figure 6.2. Location of exploratory boreholes and general geological cross-section.

power house area as a function of weathering: Region 1 -- decomposed rock; Region 2 -- weathered rock; and Region 3 -- unweathered rock (see Figure 6.2).

Considerable groundwater problems were anticipated. Artesian water conditions were encountered in several boreholes and large flows continued even throughout the dry season. Flows of 30 liters/min in the exploratory adit per probehole were typical and total water inflows between 70 to 250 liters/min have been recorded. From pressure and flow measurements taken at selected boreholes, it was calculated that water pressures reaching about 43% of the overburden rock stresses could be expected. Piezometers were installed to help with identification of potentially hazardous zones of high pore water pressures at depth.

An earthquake with a recorded magnitude of 6.3 on the Richter scale occurred in 1969 with the epicenter some 15 km from the site. Although the faults adjacent to the site appeared inactive, the possibility of renewed seismic activity was not overlooked. It was envisaged that a horizontal earthquake loading of 0.15g should be incorporated in the design of the underground power station. This was based on the author's visit to the Tongariro scheme in New Zealand where an underground power station so designed operates successfully in earthquake zones.

The extent of the exploratory geological drilling is evident from Figure 6.2 and the number of inclined boreholes should be noted due to the predominant vertical orientation of many geologic discontinuities. A total of 5,222 m of exploratory boreholes drilled from the surface involved up to four drilling rigs enabling about 200 m of core to be drilled monthly. Moreover, an underground drilling program from the test adits provided a further 550 m of core. All this drilling involved double-tube diamond core barrels to ensure high core recovery and reliable core logging. In addition to geological drilling, a substantial amount of drilling was undertaken for rock engineering investigations: 952 m of diamond drilling and 1,387 m of percussion drilling.

The geological investigations also included studies of published information about the geological history of the region, air-photo interpretation, surface mapping, water pressure testing in boreholes, as well as engineering geological mapping as construction proceeded. Geophysical surveys were also conducted in exploratory excavations.

6.2.3 *Rock engineering program*

Three types of design problems were identified at this project:

1. Excavation and reinforcement of the large span (22 m) for the power house-unprecedented in the accumulated experience from case histories in similar rock conditions;

2. Outlet (tailrace) tunnels in weathered rock; and

3. Concrete lining (instead of steel lining) of the penstocks.

Accordingly, pilot tunnels were constructed to include test adits and three full size

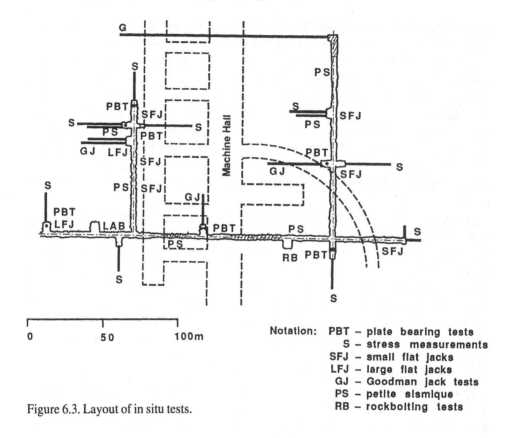

Figure 6.3. Layout of in situ tests.

Notation: PBT – plate bearing tests
              S – stress measurements
            SFJ – small flat jacks
            LFJ – large flat jacks
             GJ – Goodman jack tests
             PS – petite sismique
             RB – rockbolting tests

test excavations: a trial enlargement in the machine hall area, an access shaft and an outlet tunnel test enlargement, as well as a penstock trial chamber and test adit.

The sites for rock mechanics in situ tests were carefully selected and are depicted in Figure 6.3. The test program aimed at providing the following information for design purposes:

1. Determination of the in-situ state of stress;
2. Determination of the in-situ properties of rock masses;
3. Determination of the stress redistributions due to excavation and of the requirements for rock stability.

The rock mechanics testing program, which meets the above objectives, is presented in Tables 6.3 and 6.4. Three categories of activities may be noted: (1) field testing; (2) in-situ testing; (3) laboratory testing. These activities will now be discussed in more detail.

Field tests. The purpose of geotechnical field surveys and integral sampling (a technique enabling 100% core recovery in poor rock) was to define in detail the structural geology at all test sites and to provide engineering geological input data for rock mass classification. In addition, two field tests for rapid assessment of

Table 6.3. Observations and measurements for rock engineering characterization on site.

Property or data	Rock material	Rock mass	In-situ stress field	Modulus of deformation	Empirical design data
Field testing Geotechnical surveys and integral sampling	Detailed engineering geological description of rock strata				Input data for engineering classification of rock masses
Point-load test	Strength index from rock pieces				
Direct sheer test-portable shear box		Friction and cohesion of joints from cores			
In situ testing Overcoring cells & small flat jacks			Magnitude & directions of stresses	Deformation parameters	
Plate bearing tests & bore-hole jack		Effect of joints on strength of rock mass		Deformation parameters	
Seismic/sonic measurements (petite sismique)	Sonic velocity data on laboratory rock specimens			Longitudinal and shear wave velocities and dynamic moduli	
Convergence monitoring & borehole extensometers			Stress redistribution	Time-dependent rock mass movements around excavations	
Piezometers in boreholes		Data on water inflow, pressure and permeability of rock mass			
Rockbolt and other support trials; pullout tests					Rock support data: spacing, length, etc.

certain parameters from rock cores were used: the point-load strength index test and a direct shear test. Both these field tests involved NX size (55 mm diameter) rock cores from the localities of the in-situ tests and they utilized portable testing apparatus. The point-load index estimated the strength of the rock material while the direct shear test determined the cohesion and friction parameters along joints and bedding planes.

Table 6.4. Scope of laboratory testing for design data.

Tests	Properties	Purpose
Uniaxial compression	Uniaxial compressive strength and strength anisotropy	Elastic modulus and Poisson's ratio of rock material
'Brazilian' tensile strength	Uniaxial tensile strength	Data for tensile region of Mohr's envelope
Triaxial compression	Friction and cohesion of rock material and Mohr's envelope	Failure criterion
Physical properties	Density, porosity, water content, swelling and slake durability indexes	Weatherability and water content influence on material strength

In-situ tests. The object of the in-situ tests was to provide rock mass properties in engineering terms, namely:

1. State of stress in the rock mass before and after excavation;
2. Rock mass deformability characteristics, in different loading directions and under short and ong time duration;
3. Rock mass permeability and water conditions; and
4. Empirical data for the design of rock reinforcement.

An important consideration in conducting the in situ tests was cross-checking the results by at least two different methods. This is essential because some in situ tests can provide misleading results by being method-dependent. Accordingly, just because an in situ test is conducted does not mean that reliable results are obtained; sometimes they raise more questions than answers!

On this project, rock stress measurements were checked by two methods. The magnitude and direction of the virgin in situ stress field was determined at ten test locations using the overcoring technique known as the CSIR triaxial strain cell. At each borehole in the localities shown in Figure 6.3, at least two successful overcoring measurements were made to cross-check the results. In addition, the 'small' flat jack technique was used both to check the in situ stresses as well as determine the stress redistributions due to the excavation profile.

For rock mass properties determination, plate bearing tests were used at ten localities to determine the modulus of deformation of rock masses and their creep characteristics. The tests were made in the horizontal and vertical directions to assess rock mass anisotropy. The results were cross-checked by means of the Goodman jack technique and by the 'large' flat jack method (Bieniawski, 1978).

Seismic tests, involving the *petite sismique* technique (Bieniawski, 1980) were used as a check in determining the moduli of deformation of the rock mass with particular reference to their variability from one test location to another. This seismic technique provides velocity profiles along the drifts between the various test locations which insured that deformability of the rock mass was known throughout the exploratory tunnels and not only at the ten test locations where the plate bearing tests were conducted. The *petite sismique* technique provided measurements of both the longitudinal and the shear velocities across the excavation walls as well as in boreholes for down-the-borehole and cross-borehole surveys. These seismic measurements also provided the seismic modulus reduction ratio when compared with laboratory sonic tests on cores. This ratio was a useful index to serve as a cross-check when extrapolating rock mass conditions from the test enlargements to other localities in the underground excavation complex (Bieniawski, 1979).

Convergence measurements and multi-point borehole extensometers were used to evaluate the long term stability of the excavations, the extent of the loosened and fractured zones and to check on the displacements predicted by analyses using the finite element method.

Piezometers were installed to determine groundwater conditions in the rock mass during the wet and dry seasons and, in particular, to define the hydrological boundaries in terms of water pressure and flow. In total, 15 electric piezometers were installed in surface boreholes and 12 underground: 9 in the machine hall area and 3 in the outlet (tailrace) test enlargement.

Finally, a series of rock bolt, shotcrete and blasting trials were scheduled to provide certain empirical design data. Instrumented rock bolts of the 'Williams' type (hollow tube system) were selected to determine bolt load and deformation of the bolts which could be mechanically tensioned as well as resin grouted. The Glötzl cells were used in shotcrete to assess the rock pressure distribution around an excavation and enable effective positioning of the rock bolts and appropriate thickness of the shotcrete lining. A large radial press technique, in preference to the expensive pressure chamber test, was selected for providing design data for the lining of the penstocks. All the above in situ tests required a substantial amount of instrumentation which, in turn, required a considerable amount of drilling and core recovery. Specifications were prepared for the required underground drilling (which supplemented geological drilling), involving 952 m of diamond drilling and 1,387 m of percussion drilling (Bieniawski, 1976).

Laboratory tests. An on-site laboratory was set up near the outlet shaft and equipped to perform most standard tests. Additional facilities were available off-site at a research laboratory for specialized testing. The laboratory tests were to provide information on the behavior of rock materials as opposed to the behavior of rock masses assessed by in-situ tests. The tests aimed, where possible, at establishing a correlation between the laboratory and the in-situ test data. The following laboratory tests were conducted for this project, all in accordance with the ISRM standards (Brown, 1981):

1. *Strength and deformability in uniaxial compression.* The uniaxial compressive strength, which provides the engineer with a 'feel' for the rock material, was a parameter required for rock mass classification and yielded the upper strength limit of the rock mass. This strength was correlated with the point-load strength index to establish its reliability. The modulus of elasticity and Poisson's ratio were the necessary parameters to the analysis of the stress distributions around under-

ground excavations. As a check for anisotropic behavior, directional variations in strength and moduli were determined by compression tests at various angles to the weakness planes.

2. *Triaxial compression tests* were conducted to measure the increase in the compressive strength of rock when subjected to lateral confining pressure and to provide the cohesion and friction estimates from Mohr's envelopes for intact rock materials. The purpose here was to establish a workable strength criterion appropriate for the main rock materials found at the site.

3. *Uniaxial tensile tests* were performed to provide the information for completion of Mohr's envelopes in the tensile region.

4. *Density, porosity and water content.* Determination of rock density was necessary for estimating the rock overburden pressure while changes in rock density and porosity within the same lithological unit signified changes in the degree of weathering of the rock mass. Knowledge of porosity and water content were necessary because they affected the strength and deformation characteristics of the rock.

5. *Special tests.* These included sonic velocity tests (required for interpretation of in-situ seismic data), swelling and slake durability tests (to assess weatherability, i.e. susceptibility to disintegration under cycles of wetting and drying) and finally, petrographic analyses to provide the general mineralogy of the rock and to identify any clay minerals present (an earlier analysis revealed the presence of illite, the potential swelling of which was a concern because it could lead to difficult soft ground tunneling).

The results of the laboratory tests are summarized in Table 6.5.

Table 6.5. Results of laboratory tests on graywacke and phyllite.

	Graywacke	Phyllite	No. of tests
Density	2680 kg/m^3	2780 kg/m^3	32
Uniaxial compressive strength	148 MPa (67-228 MPa)	57 MPa (24-93 MPa)	97
Dry/wet strength ratio	1.4:1		
Triaxial compression test:			
cohesion	32 MPa	–	28
internal friction	42°	–	
Tangent modulus of elasticity at 25%σ_c	71 GPa (66-78 GPa)	56 GPa (46-69 GPa)	25
Tangent Poisson's ratio	0.25	0.27	
Sonic wave velocity	6021 m/s	–	24
Dynamic modulus	83.9 GPa (74.7-95.5 GPa)	–	
Tensile strength	17.8 MPa (11.1-25.6 MPa)	–	20
Anisotropy index (from point load test)	1.75	–	
Swelling index	0.032%	0.195%	8
Slake durability index	99.74%	97.86%	8
Shear strength (residual friction)	29°	25°	24

6.2.4 *Design investigations*

The design investigations for this scheme, falling under the direction of the author, included the overall planning of the site investigation program, performing rock mechanics tests in the laboratory and in the field, assessment of rock mass quality, interpretation of the test results, provision of the design parameters for tunnels and chambers, and design recommendations for the spacing, shape, reinforcement and construction sequence of the various excavations. In addition, close liaison had to be kept with the client, the consulting structural engineer, the consulting geologist and the site manager.

Monthly design meetings were held during which progress was reported by the leaders of these parties. In addition, reports of investigations were submitted as they were completed and it was required of all the groups involved to study these reports and submit their comments. Moreover, a detailed study was undertaken of other underground power schemes which could be relevant to this project. Not only were literature searches made but also the construction sites of numerous projects were visited and specialized design conferences attended. Visits included the Snowy Mountains Scheme in Australia, the Gordon Scheme in Tasmania, the Tongariro Project in New Zealand, the Dinorwic Scheme in Wales, the Waldeck II Scheme in Germany and the Eisenhower Tunnel in the United States. The design team, led by the author, comprised 14 engineers and technicians with these responsibilities: stress measurements (2), groundwater studies (1), plate bearing tests (2), outlet (tailrace) tests (2), laboratory tests (2), engineering geology (1), finite element analyses (1), miscellaneous testing (1), site liaison (1) and technical information (1). Administrative assistance was provided by one secretary. A radio contact was maintained between the site and the design office which was located in a city over 1500 km away. A considerable amount of travel was also necessary.

It is useful to review some of the results of the design investigations, to show how the test results were used in design decisions and to comment on the costs and benefits involved.

Concerning the results of the design investigations, once rock mass classifications were performed, a rock mass rating was found of RMR = 66 to 80 with an average of 75 (Class II: good rock). The phyllite rock mass was of Class III (fair rock) with RMR = 43 to 60 (average: 57). The Q-index for the graywacke rock was 30, ranging from 18 to 35.

It should be noted that this unique project involved some 200 *in situ* test results! In fact, two methods of stress measurements and seven different techniques for determining rock mass deformability were used, namely:

a) Rock stress measurements: 1. Overcoring technique (13 measurements), 2. Flat jack technique (10 measurements); b) Rock mass deformability: 1. Plate bearing tests (ten locations, 12 tests, 33 results), 2. 'Large' flat jacks (3 results), 3. 'Small' flat jacks (10 results), 4. Goodman borehole jack (39 results), 5. Tunnel convergence (23 results), 6. 'Petite sismique' (43 results), 7. Quality index RMR (55 results).

It was found that the major principal stress $\sigma_1 = 13.8$ MPa was about parallel to the mountain slope while $\sigma_3 = 3.7$ MPa was near-perpendicular to the slope. The intermediate stress $\sigma_2 = 5.2$ MPa was found to act in the horizontal direction parallel to the long axis of the main cavern. The overcoring technique gave the horizontal to vertical stress ratio as 0.68 to 1.18 while the flat jack results were 1.28 to 1.46. Moreover, the measured vertical stresses by both methods were found to be about twice the calculated overburden pressure. The average stress measurement results obtained at the site by the two methods did not fit the trends observed at other projects in the region and was thought to be of tectonic origin as the influence of topography could not be so pronounced.

To determine the in-situ modulus of deformation of the graywacke and phyllite rock masses, two primary testing methods were used, namely, plate bearing tests and small flat jacks. Other testing methods were also applied as part of a research program. The results from the tests are summarized in Table 6.6.

When the results of all the deformability tests were correlated with the rock mass quality RMR at the test sites (see Figure 6.4), the following overall relationship was apparent:

$$E_M = 1.76\,RMR - 84.3$$

where E_M = in situ modulus of deformation, in GPa; RMR = rock mass rating in accordance with Bieniawski (1979).

Table 6.6. Results of deformability tests at Elandsberg.

Test method	Number of tests	Range GPa	Mean GPa	Std. dev. GPa
Graywacke				
Plate bearing tests	12	34.2-58.2	44.0	13.3
Small flat jacks	10	31.7-63.9	45.5	9.4
Large flat jacks	3	34.0-56.0	42.2	8.9
Goodman jack	39	17.3-35.5	30.8	10.9
Tunnel relaxation	23	38.7-48.2	42.5	18.2
Petite sismique	41	15.5-43.1	26.0	11.6
Average in situ value (used for design)			$E_M = 40.1$	14.1
Seismic refraction (E_d)	43	30.6-91.2	65.0	27.7
Laboratory tests:				
static	32	66.9-77.9	$E_L = 73.4$	3.8 $E_M E_L = 0.59$
sonic	23	74.7-95.5	83.9	10.4
Phyllite				
Small flat jack	9	25.2-47.9	31.8	6.9
Goodman jack	6	6.0-20.0	12.0	6.2
Tunnel relaxation	4	9.7-39.6	20.0	13.4
Petite sismique	25	12.3-21.5	15.4	4.6
Average in situ value (used for design)			$E_M = 19.8$	6.5
Laboratory tests	7	46.0-69.0	$E_L = 56.0$	11.9 $E_M/E_L = 0.35$

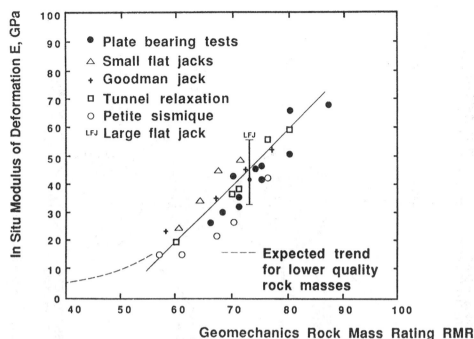

Figure 6.4. Experimental data for the rock mass modulus of deformation from various in situ tests (after Bieniawski, 1978).

This empirical equation had a correlation coefficient of 0.9612 and yielded a prediction error of 17.8% which is defined as the difference between the observed value and the predicted value expressed as a percentage of the predicted value. In view of the high correlation, the coefficients in the above equation can be rounded to

$$E_M = 2\,RMR - 100$$

which had a prediction error of 18.2%. This is a simple equation to remember and sufficiently accurate for practical engineering purposes to provide large savings in the high costs of in situ tests! Indeed, this equation was subsequently tested on 22 case histories from various parts of the world and enabled prediction of the in situ modulus of deformation to within 20%! This is a very significant finding considering that even in an extensive (and expensive!) in situ test program in fairly uniform and good quality rock mass conditions, deformability data may feature a standard deviation of 30% or as much as 14 GPa for an average modulus of 40 GPa (Bieniawski, 1984).

The choice of a design modulus for the graywacke rock mass at Elandsberg was based on the above equations and was cross-checked by using the relation E_M/E_L = 0.59 (range 0.47-0.79) derived from Table 6.6. For phyllite, the design modulus of 19.8 MPa (± 6.5 MPa) was used, with E_M/E_L = 0.35 (see Table 6.6).

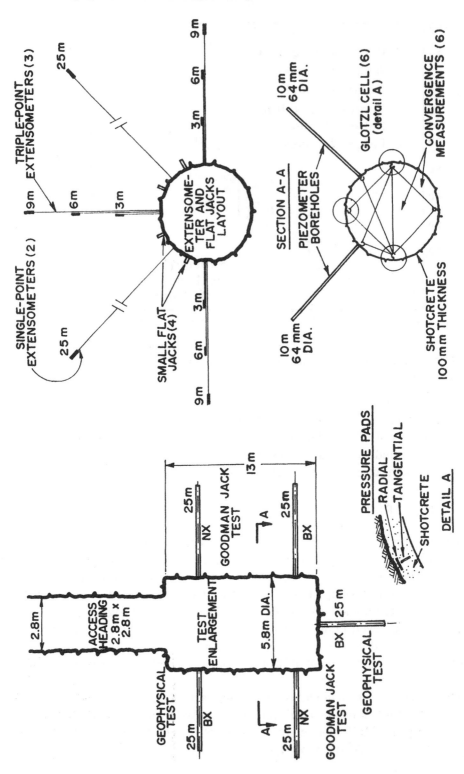

Figure 6.5. Rock engineering investigations in the outlet tunnel test enlargement (after Bieniawski, 1976).

Three specific design investigations require particular mention: outlet (tailrace) tunnel design enlargement, layout stability analysis, and sequencing of excavations in the machine hall enlargement.

1. *Outlet tunnel design study.* A full scale test excavation was constructed to verify the design solutions envisioned for the outlet (tailrace) tunnel (Bieniawski, 1984). As illustrated in Figure 6.8, the tunnel enlargement was instrumented with multiple-position borehole extensometers, convergence measuring devices, load cells on the rock bolts, Glötzl pressure pads in shotcrete, and piezometers. Rock stress measurements were carried out using two methods: (1) overcoring, featuring the CSIR triaxial strain cell, and (2) circular flat jacks. Modulus of deformation measurements were conducted using the circular flat jacks (they served two purposes), the Goodman borehole jack, tunnel convergence back-analysis and the *petite sismique* technique. Detailed engineering geological mapping and rock mass classifications were performed and geotechnical core logging was conducted on core from the instrumentation boreholes and from boreholes drilled from the surface. Groundwater conditions were assessed by a network of piezometers and by water pressure testing in boreholes.

With the in situ stress field and the rock mass modulus of deformation known, a comparison could be made between the measured parameters and those predicted by numerical analyses featuring two variations of the finite element method. This test enlargement was located in a phyllite rock mass, 102 m below the surface, and was 13 m long and between 5.8 to 6 m in diameter. It was excavated in two stages, by top heading and bench method, using the smooth blasting technique of post-splitting. The top heading was excavated first, after which all the instruments were installed. Prior to excavation of the bench, boreholes were drilled for Goodman jack tests as well as for *petite sismique* tests in boreholes. The bench was then excavated. As mentioned earlier, the phyllite rock mass was of a fair quality in accordance with the RMR classification system, having an average RMR = 57 (range 43-60). The RQD ranged from 65%-75%. The dominant discontinuity set was a near vertical foliation striking perpendicular to the axis of the test enlargement. In addition, two joint sets were present striking normal to foliation. The joints were typically tight with slightly wavy surfaces. The excavation was generally dry.

A test section, 2.5 m wide, was selected for instrumentation purposes in the middle of the test enlargement shown in Figure 6.5. A shotcrete lining, up to 200 mm thick, was applied in stages to cover the Glötzl cells but no rock bolts were installed in the test section so as not to influence the instrumentation. Stress measurements showed that the vertical stress component was 5.5 MPa while the horizontal component was 3.5 MPa. As mentioned earlier, the modulus of deformation of the phyllite rock mass was selected from Table 6.6 as 19.8 GPa with a standard deviation of 6.5 MPa.

The excavation sequence and rock mass behavior were modeled by two finite element codes: a jointed rock model and a linear elastic model (Bieniawski, 1984). The measured rock displacements and pressures were compared to predictions by these models.

To check on the extent of the damaged (loosened) rock zone around the tunnel, analyses were conducted using the Hoek-Brown criterion (Nicholson and Bieniawski, 1990). The jointed rock finite element model predicted a zone of loosening from 0.5 to 2 m. This finding was compared with the data from the seismic refraction surveys which indicated a possible zone of loosening between 2-3 m around the test enlargement which was also confirmed by the triple-point extensometers. The elastic finite element method predicted a much smaller damaged rock zone.

It was concluded from these investigations that these results were satisfactory and the proposed design of the outlet (tailrace) tunnel was fully acceptable as safe and economical.

2. *Stability analyses of the layout of excavations.* Finite element analyses were performed to assess the merits of alternative layouts of underground excavations. Based on the in situ tests, ranges of values for the modulus of elasticity and the applied stresses were used. It was decided that the main value of numerical analyses was in comparative studies rather than in absolute predictions. Accordingly, the rock mass was assumed to behave in an isotropic, linear elastic manner and two-dimensional analyses were selected.

The objective was to determine the stresses and displacements around the three main underground caverns, the transformer hall, the machine chamber, and the valve chamber, as follows:

1. Comparison of a three-cavern layout with other arrangements;

2. Determination of the influence of the cavern spacing;

3. Determination of the influence of the rock modulus on the magnitude of the displacement;

4. Determination of the influence of the ratio (k) of the horizontal to the vertical stress components.

The results indicated that the maximum stress around the caverns was about 30 MPa and the maximum displacement of the sidewalls about 25 mm.

For horizontal to vertical stress ratios (k) of unity, the distances between the cavern center lines should ideally be approximately 48 m between the transformer hall and the machine hall, and 53 m between the machine hall and the valve chamber. If the caverns were spaced closer together, large volumes of rock between these excavations would be subjected to tensile stresses.

Increases in the ratio k resulted in considerable increases in the magnitude of stresses at points of high stress concentration and in the displacements of the cavern sidewalls. For example, the maximum compressive stress in the roof of the machine hall was found to increase from 21 MPa (k = 0.5) to 63 MPa (k = 3). The maximum displacement at the right hand wall of the machine hall was found to increase from 7.4 mm (k = 0.5) to 55 mm (k = 3), for a rock modulus of 20 GPa. In addition, the zones of tensile stress increased considerably with the increasing stress ratio. As a result, since the actual stress ratio of up to 1.5 was determined by in situ stress measurements, the distance between the caverns was increased and their shapes changed to reduce stress concentrations.

3. *Sequencing of excavations in machine hall enlargement.* Due to the very large excavation span necessary for the machine hall (22 m) with no precedent available in similar rock conditions, and due to design changes in the shape of the

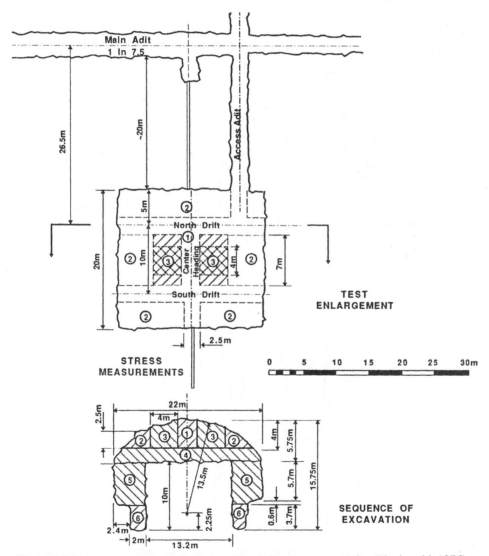

Figure 6.6. Excavation sequence in the machine hall enlargement (after Bieniawski, 1976).

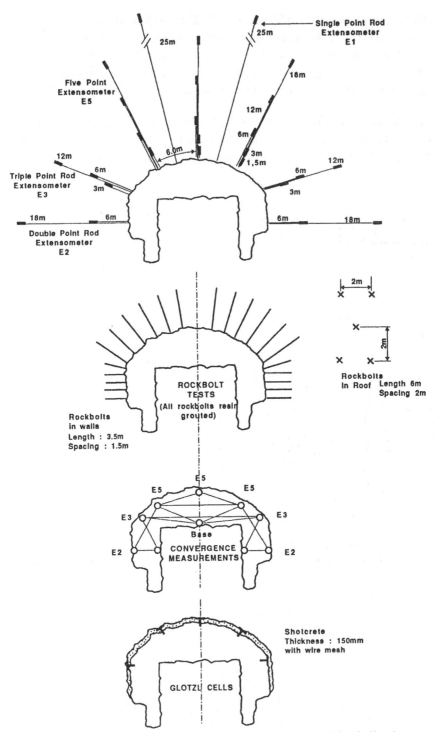

Figure 6.7. Rock reinforcement and instrumentation in the machine hall enlargement (after Bieniawski, 1976).

roof arising from the above stability studies, the design of the rock reinforcement in the main cavern had to be checked by actual field trials. In addition, in those days, cavern support by rock bolts and shotcrete was somewhat of a rarity. Standard practice was a reinforced concrete arch. For the Elandsberg project a cavern reinforced by rock bolts and shotcrete was designed because a concrete arch would add another 3 m to the span. It was necessary to reassure the client that this innovative design was as stable as it was economical.

Accordingly, a full scale test enlargement was planned for the machine hall featuring not only a series of in situ tests but also a carefully monitored excavation sequence accompanied by an evaluation of rock mass stability following installation of rock reinforcement in the form of rock bolts and shotcrete.

Figures 6.6 and 6.7 present the design of the excavation sequence, rock reinforcement and instrumentation for the machine hall chamber (Bieniawski, 1976).

The shape of the excavation cross-section should be noted particularly the long 'ears' at the side walls. Their purpose was to release lateral support provided by the rock to the sidewalls and thus enable the development of full displacements as would be the case once the actual cavern was excavated.

The remarkable aspect of this design study was that it was planned in 1976 and its scope and methodology have not been matched even to this day! Moreover, as will be seen below, cost and labor force were very modest by present standards. This should be a sobering thought when considering Design Case 3 representing the current situation.

6.2.5 *Costs and benefits of the employed design methodology*

The design methodology employed at Elandsberg should be examined from the point of view of the costs and benefits to the owner of the project. In view of the fact that the rock engineering investigations were so extensive in scope and so ambitious in effort, it may indeed be asked what were the direct benefits of the exploratory tunnels and enlargements and what were the costs involved in this design case. As said earlier, because the author realized the uniqueness of the project from the start, he kept a design diary recording the technical events, personal observations, design decisions, and costs. Some of this information has been published (Bieniawski, 1976; 1984) and the permission of the owner – one of the most progressive and forward looking engineering companies – has been acknowledged.

The justification for the in situ trials has already been explained and lies in the fact that there was no other reliable way than a full scale test enlargement to predict the maximum rock spans for the power house. In addition, in view of the unusual depth of weathering, i.e. 60-100 m below surface, through which the outlet (tailrace) tunnels had to pass, a tailrace test enlargement was necessary for the most economical tunnel design. Moreover, only a large in-situ test such as a radial press test could determine reliably whether a concrete lining, instead of steel, was acceptable for the penstocks. This alone saved the owner about $260,000 (in 1977 dollars).

Apart from these justifications of the exploratory excavations, there were direct benefits, namely the tunnels could be utilized when actual construction started. For example, the access tunnel to the exploratory works would be enlarged to construct the eventual main access tunnel. Most of all, the rock conditions would be so well known that the element of risk of 'unforeseen conditions' would be removed from the contractor's bid thus resulting in a substantially lower overall contract price. As far as the costs of the design investigations are concerned, the total costs including the site characterization program and exploratory tunnels were 2.8% of the civil construction costs of the project, as listed below. The cost of the rock engineering investigations alone (design and data analyses, equipment, and drilling of test boreholes in adits) was $670,600 or only 0.6% of the civil construction costs. Further cost details are listed below (in 1980 dollars):

Mechanical and electrical installations	$133,000,000 (41.6%)
Surface facilities	63,800,000 (19.9%)
Civil construction	119,700,000 (37.4%)
Design costs	3,353,000 (1.05%)
	(2.8% of civil construction)
Overall costs	$319,853,000 (100%)

Details of the design costs	
Total design costs	$3,353,000 (100%)
Pilot tunnels and test adits	1,995,000 (59.5%)
Drilling for geologic site characterization	687,400 (20.5%)
Design and data analyses (3 yrs)	368,600 (11.0%)
Rock mechanics equipment	168,000 (5.0%)
Test boreholes in adits	134,000 (4.0%)

It is interesting to recall that a report by the US National Committee on Tunneling Technology (USNCTT, 1984) recommended that 'expenditures for geotechnical site exploration should be increased to an average of 3% of the estimated project cost, for better overall results.'

It is concluded from this project that well planned, well executed design investigations were performed over 15 years ago using a range of tests and methodologies that would be hard to match even today. In fact, the author's involvement on this project provided him with such a wealth of design experience that it can be used on any modern project without fear of being out-of-date. On the contrary, some recent design investigations on major projects do not even come close to those conducted at Elandsberg so long ago. For example, no plate bearing tests of this scope have been performed in the United States in the past decade! There is certainly a multitude of numerical studies but how reliable are the input data used in these analyses? There is also an overabundance of rock mechanics laboratory tests the results of which fill thick volumes of annual rock mechanics symposia, but how is this information used in design? Alas, hardly ever!

The significance of the Elandsberg design case study is that it is an example of the best design practice due to superb teamwork, detailed planning, the large degree of latitude given by the owner to the leaders of the parties involved, and minimal red-tape. As a result, the work was accomplished within three years and on a budget so reasonable that it seams hardly possible today. Consider by way of contrast that the radioactive waste repository program in the United States has cost more than $2.5 billion since 1982 without having much to show for it (Fairweather, 1991).

To conclude, let us summarize the functional requirements and design components, appropriate to this project, in terms of the design methodology proposed in Chapter 4.

The objective was to design a large underground power house and the associated tunnels and chambers to provide 1000 MW of power derived from a pumped-storage hydroelectric plant. There were three major constraints since this was a pumped-storage hydroelectric scheme: (1) its location was fixed, (2) the orientation of the machine hall was dictated by the hydroelectric requirements, and (3) the spans of the chambers and tunnels were also fixed as per the operational and equipment needs. To satisfy the design objectives, the functional requirements were as follows:

FR_1 = assure stability of a 22 m span machine hall without a reinforced concrete arch; and

FR_2 = provide stable 6 m diameter outlet (tailrace) tunnels in weathered rock.
The design solution was characterized by:

DC_1 = cavern profiles free of significant stress concentrations, reinforced by rock bolts and shotcrete, with demonstrated stability in a full scale machine hall enlargement;

DC_2 = circular outlet (tailrace) tunnel, smooth blasted, reinforced by rock bolts and shotcrete lined, with demonstrated stability in a full scale tunnel enlargement.

6.3 DESIGN CASE STUDY 2: COAL MINING

In late 1979, a group of design consultants – working under the author's guidance – began an evaluation of the strata control measures in use at an underground coal mine located in West Virginia. The mine had been plagued with ground control problems, such as excessive roof falls and shearing of rock bolts in the roof, since its opening in 1975. The consulting team conducted a careful study of the problem, completed its report, but before the recommendations could be implemented the mine was closed down due to adverse market conditions.

The purpose of this case study is to evaluate the consulting engineers recommendations in the light of today's technology, to identify any shortcomings in the design approaches used, and to propose an improved design of the ground control system which would utilize the design methodology proposed in Chapter 4. It is assumed that the reader of this chapter has the basic knowledge of strata control in coal mines.

6.3.1 *Statement of the design problem*

The mine management informed the consulting team of five design objectives arising out of the problems encountered at the mine:

1. Eliminate progressive roof failure (known as 'cutter roof') which results in roof falls;

2. Prevent the shearing of rock bolts in the roof;

3. Maintain an unobstructed opening of a width of 18 ft (5.5 m);

4. Control the roof-to-floor convergence which must not exceed 1 ft (0.3 m); and

5. Ensure that the overall coal extraction is greater than 75%.

6.3.2 *Background information on the coal mine*

This underground coal mine has an average overburden thickness of 600 ft (183 m) with a coal seam thickness of 5.2 ft (1.6 m). The immediate roof consists of a thinly laminated shale, 1.5 ft (0.5 m) thick, and the main roof is a thick bedded shale. The floor consists of a hard clay layer, approximately 3 ft (0.9 m) in thickness.

The consulting engineers conducted a fairly thorough site investigation and data collection program, and the results are summarized below.

Laboratory tests. Laboratory tests were conducted on specimens from the immediate roof to determine the uniaxial and triaxial compression strength, density, shear strength, and indirect tensile strength. The averaged results are summarized in Table 6.7.

The consulting team also tested core and cube specimens from the coal seam as well as core specimens from the floor. The results are summarized in Tables 6.8 and 6.9, respectively.

Table 6.7. Laboratory test results: Immediate roof rock.

Parameter	Average value	Range of values
Uniaxial compressive strength	6000 psi (41 MPa)	5690-6710 psi
Internal friction angle	26.5°	
Density	175 pcf (26 kN/m^3)	155-180 pcf
Shear strength	3250 psi (22 MPa)	1845-3970 psi
Tensile strength	1350 psi (9.3 MPa)	1080-1490 psi
Modulus of elasticity	2×10^6 psi (13.8 GPa)	1.8-2.3×10^6 psi
Poisson's ratio	0.25	0.22-0.3

Table 6.8. Laboratory test results: Coal seam.

Parameter	Average value	Range of values
Uniaxial compressive strength	4300 psi (29.6 MPa)	3850-4620 psi
Indirect tensile strength	169 psi (1.2 MPa)	115-216 psi
Modulus of elasticity	5×10^6 psi (34.5 GPa)	$4.5\text{-}5.3 \times 10^6$ psi
Poisson's ratio	0.24	0.15-0.29
Density	81 pcf (12.7 kN/m^3)	75-87 pcf

Table 6.9. Laboratory test results: Floor rock.

Parameter	Average value	Range of values
Uniaxial compressive strength	1680 psi (11.6 MPa)	1566-1842 psi
Cohesion	30 psi (206 kPa)	
Internal friction angle	19°	
Density	159 pcf (25 kN/m^3)	140-162 pcf
Modulus of elasticity	0.7×10^6 psi (4.8 GPa)	$0.64\text{-}0.84 \times 10^6$ psi
Poisson's ratio	0.3	0.28-0.32

In-situ stress measurements. The overcoring technique was used to determine the magnitude and orientation of the principal stresses. The major principal stress was found to be horizontal, measured as 1900 psi (13 MPa) and oriented at N65°W. The minor principal stress of 600 psi (4 MPa) was in the vertical direction.

6.3.3 *Features of the consulting team's design*

The design procedure included floor strength analysis, roof span analysis, roof support design, and pillar design.

Floor strength analysis. An evaluation of the bearing capacity was conducted to analyze the floor strength using an equation from soil mechanics:

$$q = cN_c + \gamma D_f N_q + 0.5B\,N_\gamma \qquad (6.1)$$

where N_c, N_q, N_γ are bearing capacity factors (dependent on the internal friction angle); c = cohesion; γ = density; D_f = depth of footing (pillar punching into floor); B = width of footing (or pillar).

The ultimate bearing capacity calculated by this method was of the order of 60 psi (413 kPa), a value much smaller than the typical pillar load. This signified a bearing capacity failure based on the soil mechanics analysis performed. However, as only a very thin layer of the floor was a soil-like material, the consulting engineers reasoned that this analysis was not appropiate after all.

Roof span analysis. The consulting engineers primarily used numerical techniques to analyze failure mechanisms for various opening sizes. A finite element analysis showed that for a σ_v/σ_h ratio of $^1/_3$, a maximum span of 18 ft (5.5 m) was possible before significant tension would occur at the roof centerline. The simulation of principal stress directions around the opening suggested a possible progressive failure above the ribline.

Roof support design. The recommended roof bolting plan consisted of 36 in (0.9 m) long bolts on a 4 × 5 ft (1.2 m × 1.5 m) pattern. Mechanical type rock bolts were proposed with a ⅝ inch (16 mm) diameter and grade 40 steel. The yield load of this bolt is 14,000 lbs (6.3 tonnes). The anchorage capacity of the bolt was found to be greater than the yield load, therefore bolt tension was calculated as 8400 lbs (3.8 tonnes) (60% of yield load). The bolt length was selected on the basis of geologic investigations, which revealed an 18 inch (0.45 m) thick immediate roof. Based on practical considerations, the design team selected a 4 × 5 ft (1.2 m × 1.5 m) spacing pattern of rock bolts.

To control the 'cutter roof' failure at the ribline, the consulting team proposed to install 7 ft (2.1 m) resin bolts angled at 20° from the horizontal spaced every 10 ft (3 m) along the ribline. The roof bolting layout is shown in Figure 6.8. The 20° angle was chosen to align the bolts perpendicular to the major compressive forces determined from finite element analysis.

Pillar design. For pillar design purposes, the in situ strength of coal was estimated by conducting a regression analysis on laboratory strength data of various cube sizes to derive the strength of a 60-in (1.5 m) specimen. This resulted in an in situ coal strength of 588 psi (4 MPa).

Pillar load was calculated by the tributary area method, while the overall pillar strength was established by comparing four formulas: those proposed by Obert-Duvall, Holland-Gaddy, Salamon-Munro, and Bieniawski. The pillar strength for a range of pillar sizes was calculated using each method and the safety factor was

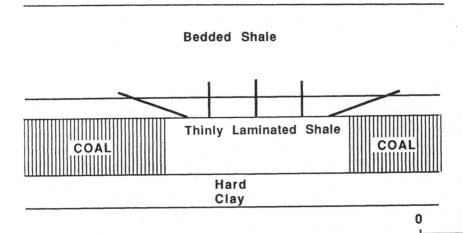

Figure 6.8. Section view of the consulting engineers roof bolting layout for a coal mine (after Kicker, 1990).

determined for each pillar size knowing the pillar load from the tributary area approach.

The pillar sizes selected by the consulting team were to facilitate retreat mining and they are listed in Table 6.10 and shown in Figure 6.9.

Critique. The consulting team's design of the underground coal mine had a number of strong points. They performed a fairly extensive site investigation and collected useful data. As a result of the in situ stress measurements, they recommended that the direction of the mains be reoriented to a direction parallel to the major horizontal stress. Finite element studies were conducted which confirmed the potential failure areas above the ribline and provided the orientation of the proposed angled bolts.

There were, however, some weaknesses in the proposed design, as viewed today. It is generally believed that to form the necessary compressive arch across

Table 6.10. Pillar sizes proposed by the consulting team.

Pillar type	Pillar size	Safety factors
Main entries	60 × 80 ft (18 × 24 m)	2.1 – 3.1
Panels	60 × 90 ft (20 × 27.5 m)	2.5 – 3.7

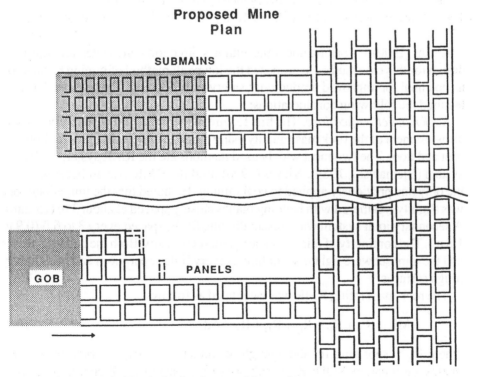

Figure 6.9. Proposed pillar design by the consulting engineers.

an opening using tensioned mechanical bolts, the bolt length-to-spacing ratio should be greater that 1.2. According to the consulting team's design, the length-to-spacing ratio was less than unity. Also, the estimated in situ coal strength used in the pillar design formulas was quite low. This led to a considerable over-design of the pillars under normal stress conditions. However, in the pillar retreat mining sequence proposed, abutment stresses would develop along the panels. The consulting team did not treat these abutment stresses in their design, but as it turned out, because of the over-design stated above, the error was somewhat canceled out.

The consulting team recommended this room-and-pillar retreat approach in 1980, during a time when longwall mining was a relatively new practice. In today's technology, longwall mining would be a much more efficient choice (Bieniawski, 1987).

6.3.4 *Improved design*

In accordance with the concepts proposed in Chapter 4, there are three functional requirements in this case study:

FR_1 = Eliminate progressive roof failure ('cutter roof');
FR_2 = Prevent the shearing of roof bolts; and
FR_3 = Control roof to floor convergence to below 1 ft (0.3 m).

The constraints are: an entry width of 18 ft (5.5 m) and minimum coal extraction of 75%.

Kicker (1990) – working under the author's direction – presented an improved design for this case study. He proposed the longwall method of mining because of its potential for improved productivity, safety and economy. The improved design featured the following considerations:

Rock mass classifications. The first feature in the improved design process was an assessment of the rock mass quality using classification techniques for both the roof and floor. The two most widely used methods are the RMR system and the Q system (Bieniawski, 1989). Kicker (1990) used the RMR system because it was the only one previously applied to coal mining. He noted that the immediate roof was slightly weathered with bedding planes closely spaced at about 3 in (75 mm). A secondary joint system was present dipping 85°E, spaced every 3 to 6 ft (0.9 to 1.8 m). The continuity of this secondary joint set was low. Both the bedding planes and the joint set were tight discontinuities (very little separation) and had slightly rough surfaces.

The rock mass rating RMR was calculated as follows:

$$RMR = R_A + R_B + R_C + R_D + R_E - Adj \tag{6.2}$$

where R_A = subrating for the strength of intact rock; R_B = subrating for rock quality designation RQD; R_C = subrating for spacing of discontinuties; R_D =

Table 6.11. Rock mass quality ratings for roof and floor strata.

	R_A	R_B	R_C	R_D	R_E	Adj	RMR
Roof rock	5	16	5	20	15	–5	56
Floor rock	2	7	8	12	15	–5	39

subrating for conditions of discontinuties; R_E = subrating for groundwater; Adj = adjustment for orientation of discontinuties.

The calculated values of the ratings are shown in Table 6.11 giving RMR = 56 for the roof strata and RMR = 39 for the floor rock.

Roof support design. The following procedure was used by Kicker (1990) to determine roof support requirements:

1. The rock load height was based on the RMR and the opening width, B:

$$h_t = [(100\text{-RMR})/100] * B = 7.9 \text{ ft } (2.4 \text{ m})$$ (6.3)

2. The bolt length was calculated as:

$$L = h_t / 2 = 4 \text{ ft } (1.2 \text{ m})$$ (6.4)

and

$$L = B^{2/3} = 18^{2/3} = 6 \text{ ft } (1.8 \text{ m})$$ (6.5)

3. The bolt capacity was considered as the lesser of the anchorage failure load and the bolt yield load:

$$L_{anchor} = 9.5 \text{ tons (from field pull-out tests)};$$

$$L_{yield} = 8.8 \text{ tons (for } ^3\!/_4 \text{ inch (19 mm) dia., grade 40 steel)};$$

Therefore, the bolt capacity

$$C_b = 8.8 \text{ tons.}$$

4. The bolt spacing was based on the criterion:

$$L / S > 1.2$$

where L = bolt length and S = bolt spacing.

Because of the relatively short term use of panel entries and because of the roof rock being of good quality, mechanical bolts were recommended. Grade 40 steel with a $^3\!/_4$ inch (19 mm) diameter provides an adequate yield strength which is slightly less than the bolt anchorage capacity.

Based on the above analysis, a bolt length of 4 ft (1.2 m) and bolt spacing of 3 ft (0.9 m) would be one possible solution, providing an adequate bolt length according to the RMR and satisfying the length-to-spacing ratio criterion. The 7 ft (2.1 m) resin bolts angled at 20° from the horizontal, as recommended by the consulting team in 1980, would still be a valid means for control of the 'cutter

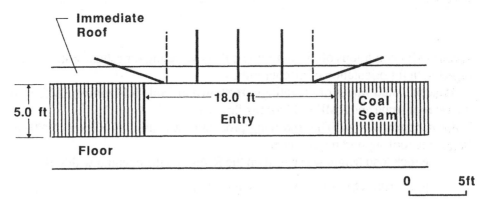

Figure 6.10. Alternative solution to roof bolting: section view (after Kicker, 1990).

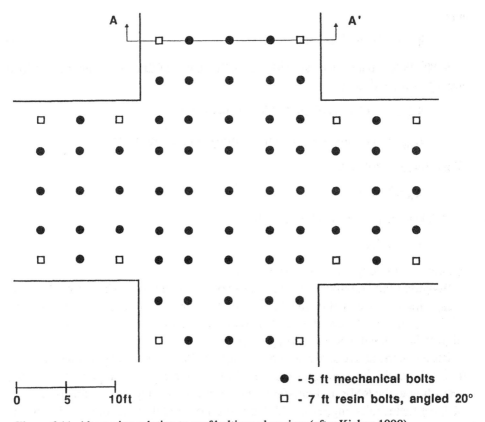

Figure 6.11. Alternative solution to roof bolting: plan view (after Kicker, 1990).

roof' failure. Resin grouting of the angled bolts would be preferred because of a higher resistance to shear failure.

An alternative solution would be to reduce the number, or density, of bolts while still maintaining the proper bolt length-to-spacing ratio. This would feature 5 ft (1.5 m) bolts on 4 ft (1.2 m) spacing as shown in Figures 6.10 and 6.11. This arrangement, being more economical, is therefore the recommended solution.

Pillar design. For pillar design under development loading, five formulas were used by Kicker (1990) to estimate the strength of coal pillars. Calculating the pillar stress based on the tributary area approach and applying the appropriate safety factors, the pillar width was determined in accordance with each method. The methods are detailed in Kicker (1990). The results are summarized in Table 6.12.

As stated earlier, the longwall method was proposed as the method of mining. In this case, the abutment loading on the chain pillars is of prime interest. Four approaches to longwall chain pillar design were considered (Bieniawski, 1987): Carr and Wilson's method, Hsiung and Peng's method, Mark and Bieniawski's method (called ALPS for Analysis of Longwall Pillar Stability), and Choi and McCain's method. The resulting pillar widths are shown in Table 6.13.

Kicker (1990) provided a comparison of the results from the five design approaches and this is shown in Figure 6.12. For a meaningful comparison, each method in Figure 6.12 is presented using a safety factor of unity (with the exception of the Choi and McCain method in which there is a built-in safety factor of 1.3). A three-entry system was used with equal-size pillars except where noted. Where an abutment pillar system is used, the abutment pillar is located adjacent to the headgate, and the yield pillar width is 20 ft (6 m). Observations from this

Table 6.12. Results from pillar strength formulas (after Kicker, 1990).

Method by	Width, ft (m)	Factor of safety
Obert-Duvall	26 (7.9 m)	2.0
Holland-Gaddy	42 (12.8 m)	2.0
Holland	32 (9.8 m)	2.0
Salamon-Munro	24 (7.3 m)	1.6
Bieniawski	24 (7.3 m)	1.5

Table 6.13. Comparison of chain pillar widths by alternative design methods (after Kicker, 1990).

Method by	Width × length, ft (m)	Factor of safety
Carr and Wilson	50 × 50 (15.3 × 15.3 m)	1.0
Hsiung and Peng	76 × 76 (23 × 23 m)	1.0
Mark and Bieniawski	43 × 43 (13 × 13 m)	1.3
abutment	54 × 65 (16.5 × 20 m)	1.3
yield pillar	20 × 65 (6.1 × 20 m)	1.0
Choi and McCain	28 × 80 (8.5 × 24.4 m)	1.3

analysis reveal that Wilson's approach is very sensitive to the k value (a function of internal friction angle). It can also be seen that the Choi and McCain method (which was developed solely for abutment pillar design) becomes very conservative at depths greater that 1000 ft (305 m) and is difficult to use at depths less than 500 ft (152 m).

The ALPS method of Mark and Bieniawski (1986) has the greatest flexibility and can be used for equal-size pillars or yield-abutment pillars. This analysis suggested that the three most useful chain pillar design approaches are Carr and Wilson (with k = 3.25), Hsiung and Peng, and Mark and Bieniawski (ALPS).

When a safety factor of 1.3 is used, as recommended for the Mark and Bieniawski method, it yields mid-range pillar widths compared to alternative approaches – neither too conservative nor overly daring.

Based on the above analysis, the following pillar sizes were recommended by Kicker (1990) and are justified below:

Pillars in the main entries: 30 × 37 ft (9 m × 11 m)
Bleeder pillars: 37 × 37 ft (11 m × 11 m)
Chain pillars: abutment pillar: 54 × 65 ft (16.5 m × 20 m)
                  yield pillar: 20 × 65 ft (6 m × 20 m)
Barrier pillars: 100 × 697 ft (30 m × 212.5 m).

The chain pillars were sized according to the Mark and Bieniawski (ALPS) analysis with a safety factor of 1.3. The ALPS method yields the highest

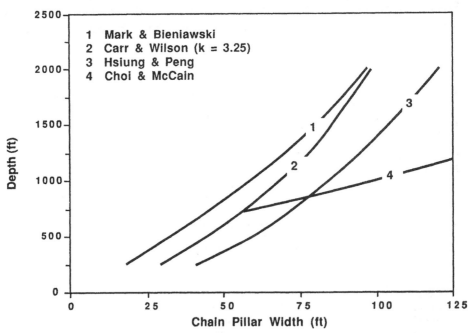

Figure 6.12. Comparison of longwall pillar design approaches (after Kicker, 1990).

extraction ratio and also uses a tested safety factor. The mains were sized based on the Bieniawski formula, with a safety factor of 2.0. Using a safety factor of 1.3, the ALPS method recommends a bleeder pillar size of 39 ft (12 m). However, because of the protection from abutment stresses by the barrier pillar, a slightly smaller safety factor is acceptable, therefore a pillar width of 37 ft (11 m) was proposed. The standard barrier pillar formulas suggested a pillar width of 100 ft (30 m) based on a width-to-height ratio of 20. The exact panel dimensions are therefore 697 × 6954 ft (212.5 m × 2121 m). A schematic layout of this design is provided in Figure 6.13. The panel area is approximately 76% of the total area,

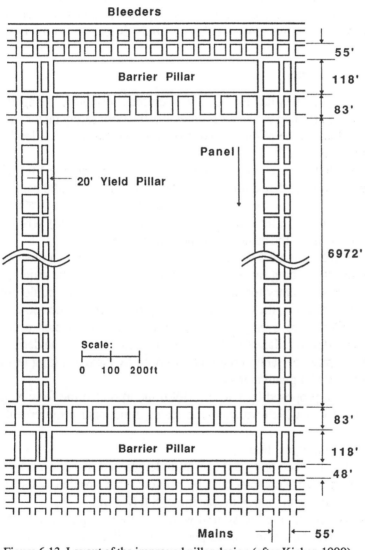

Figure 6.13. Layout of the improved pillar design (after Kicker, 1990).

with 100% coal extraction. The development extraction ratio is about 0.46. Therefore, the total extraction ratio is approximately e = (1)(0.76) + (0.46)(0.24) = 0.87 or the overall coal extraction is 87%.

Floor design. The bearing capacity was determined by two different methods (Kicker, 1990). The first method was described by Bieniawski (1987) and is similar to the method used by the consulting team, but modified for mine floors. The bearing capacity from this method resulted in q = 648 psi (4.5 MPa) – far below the predicted pillar stress of 2477 psi (17 MPa), as determined by the ALPS method.

Due to the unreliability of bearing capacity approaches to floor design in mines, a new method was developed by Faria and Bieniawski (1989). The Faria-Bieniawski method requires fairly extensive data from laboratory tests, which fortunately the consulting team has provided. This analysis resulted in the floor bearing capacity

$$q = 1541 \text{ psi } (10.6 \text{ MPa})$$

Note that, for floor design, safety factors of 2.0 are required for stability. Using the maximum pillar load determined by the ALPS method, the floor safety factor was 0.62. If the *average* load on the pillar is used, the safety factor is 1.16. Clearly, floor failure was to be expected.

Indeed, floor heave had occurred in the mine, at one time approaching 1 ft (0.3 m). Because it was anticipated that abutment stresses of the order of 2500 psi (17 MPa) could occur, floor control measures were necessary. It was proposed to relieve the stresses occurring in the floor by cutting a 6 inch (150 mm) kerf (or vertical slot) in the floor along the center of the headgate entry.

Summary. The consulting team's design was found to be lacking in a few key areas. The bolt length-to-spacing ratio was less than 1.2, therefore the required compressive arch across the entry formed by the tensioned rock bolts could not develop. In the sizing of pillars, the estimated in situ pillar strength was too low and therefore the pillars were over-designed. Abutment stresses, which occur along the panel entries due to pillar extraction, were not considered in the design. Finally, the room-and-pillar extraction method proposed by the consulting team was not as efficient as the longwall method.

The improved design proposed by Kicker (1990) incorporated the following design components (DCs) which satisfied the stated functional requirements (FRs):

DC$_1$: The longwall mining method.

DC$_2$: Mechanically tensioned bolts, 5 ft (1.5 m) in length and spaced on a 4 × 4 ft (1.2 m × 1.2 m) pattern, supplemented by angled resin grouted bolts, as designed by the consulting team, but with spacing of 8 ft (2.5 m).

DC$_3$: A yield-abutment chain pillar system which offered a higher extraction ratio and provided stress relief in the tailgate. Chain pillars were designed according to the Mark and Bieniawski (1986) method with an abutment pillar size of 54×65 ft (16.5 m $\times 20$ m) and a yield pillar size of 20×65 ft (6 m $\times 20$ m). Pillars in the mains were designed using the Bieniawski (1987) formula resulting in a pillar size of 30×37 ft (9 m $\times 11$ m).

DC_4: Bearing capacity analysis after Faria and Bieniawski (1989) indicated possible floor failure, therefore the stress relief technique of cutting a kerf along the center of the headgate entry was incorporated.

In accordance with the concepts of design methodology proposed in Chapter 4, the three functional requirements FR_1 through FR_3 are satisfied by the corresponding three design components DC_2 through DC_4. Design component DC_1 provides compliance with the two constraints.

Moreover, the comparisons of the alternative pillar design approaches were good examples of the decision-making process as well as the evaluation and optimization procedures used in this case study.

If this design were implemented, instrumentation and monitoring of the actual performance would be necessary to check the effectiveness of the selected design.

6.4 DESIGN CASE STUDY 3: RADIOACTIVE WASTE REPOSITORY

Long-term storage of high-level radioactive waste presents formidable challenges throughout the world. In the USA, the design of an underground repository for high-level nuclear waste (HLW) was mandated by a 1982 act of Congress to provide safe underground storage for a period of 10,000 years, a criterion without precedence in rock engineering! One of the major problems encountered to date was the lack of a design methodology which could contribute to the resolution of design issues arising out of complex licensing criteria and strict environmental standards.

The purpose of this case history is to demonstrate the difficulties associated with problem definition (when political considerations interfere with engineering issues) and give examples of the applicable functional requirements, constraints and design components, in accordance with the design methodology proposed in Chapter 4. Hopefully, this will assist in the design of a repository, an activity which has already had a turbulent history and still has a long way to go before radioactive waste storage scheduled for the year 2010. But first, some background information is appropriate.

6.4.1 *Background*

At present, approximately 20% of the world's electricity is derived from 426 nuclear power plants operating in 42 nations. A further 96 reactors were under construction in 1991, including four in the United States. The United States derives 19.1% of its electricity from nuclear energy but in some countries, this percentage is much higher, e.g. France with 74.6% and Sweden with 45.1%. The challenge of radioactive waste disposal derives from the spent fuel from these nuclear power plants. Each 1,000 MW nuclear power plant produces about 30 tons of spent fuel a year which, if reprocessed, could be reduced to about 10 cubic meters of highly radioactive glass. However, some countries, including the United States which has 111 nuclear reactors in operation, have chosen to dispose of commercial spent fuel directly. It is estimated that by the year 2000 American nuclear power plants will have produced some 72,000 metric tons of radioactive waste, enough to cover ten football fields one meter deep!

Over the years, many concepts have been proposed for the permanent isolation of high-level waste, including subseabed and space disposal, but there is a strong worldwide consensus that the best and safest long-term option for dealing with nuclear waste is geological isolation. This means storage of the waste in specially designed and engineered underground facilities where local geology and ground-water conditions ensure isolation of the waste for tens of thousands of years.

The United States, as the first nation to produce nuclear energy, is faced with the earliest deadline for storage. The Nuclear Waste Policy Act, passed by Congress in 1982, assigned responsibility to the Department of Energy (DOE) for designing and eventually operating a deep geologic repository for high-level radioactive waste. The repository was to be licensed by the Nuclear Regulatory Commission (NRC) and must meet radionuclide release limits, based on a generic repository, that would result in less than 1000 deaths in 10,000 years as specified in a standard established by the Environmental Protection Agency (EPA).

It was pointed out by the National Academy of Sciences (NAS, 1990) that the US program is unique in its rigid schedule, in its insistence on defining in advance the technical requirements for every part of the multibarrier system, and in its major emphasis on the geological component of the barrier as detailed in the NRC regulations. In essence, nuclear waste management is a tightly regulated activity, hedged with laws and regulations, criteria and standards. Some of these rules call for detailed predictions of rock behavior for tens of thousands of years, longer than recorded human history.

Other nations, notably Canada, Sweden and Germany follow an alternative approach characterized by flexiblity in the exploration, design and construction of a geologic test facility and a low – to medium – level waste repository. This flexible approach allows each step in the characterization and design to draw on information and understanding developed during previous steps, and from prior experience with similar construction projects (Wilson et al. 1991; Langer, 1989; Morfeldt, 1989). During storage and subsequent to the closing of the repository, emphasis will be on monitoring and on the ability to repair faulty systems, in order to minimize the possibility that unplanned or unexpected events will compromise the integrity of the disposal system.

The Board on Radioactive Waste Management of the National Academy of Sciences) (NAS, 1990) reported that the US program as conceived and implemented over the past decade, *is unlikely to succeed because it is poorly matched to the technical task at hand*. This is because the program assumes that the properties and future behavior of a geologic repository can be determined and specified with a high degree of certainty. In reality, the inherent variability of the geologic environment will necessitate frequent changes in the specifications.

Another independent body, established by Congress in 1987, is the Nuclear Waste Technical Review Board (NWTRB). Its purpose is to evaluate the technical and scientific validity of activities undertaken by the DOE and, in the process, comment on the appropriateness of NRC regulations and EPA standards. In its reports (NWTRB, 1990), this Board also expressed many concerns and made far-reaching recommendations which would significantly improve the US nuclear waste disposal program. It pointed out that the standards should not impose restrictions that would *foreclose at the outset a candidate site subsequently shown to be suitable based on sound scientific considerations*. Moreover, an urgent need was identified to develop a comprehensive methodology for performance assessment of the overall repository concept.

It should also be noted that some of the concerns reported by the National Academy of Sciences (NAS, 1990) are shared by the Board (NWTRB, 1990), however, some were based on 1988 conditions and did not take into account the positive changes that have occurred in the US nuclear waste program since April 1990 when a new director of the DOE civilian radioactive waste management office was appointed.

In addition, the new US Secretary of Energy in his report to Congress recommended an orderly program of scientific investigations, one not driven by unrealistic scheduling demands, and the establishment of a monitored retrievable storage (MRS) facility that would be *unlinked* to the repository construction schedule to allow early acceptance of spent nuclear fuel.

There is no scientific reason why a satisfactory high-level nuclear waste repository cannot be built and licensed (NAS, 1990), but an alternative approach is needed which is more flexible and experimental. It should be based on these premises (NAS, 1990):

a) Surprises are inevitable in the course of investigating any proposed site, and things are bound to go wrong on a minor scale in the development of a repository; and

b) If the repository design can be changed in response to new information, minor problems can be fixed without affecting safety. Major problems, if they appear, can be remedied before damage is done to the environment or to public health.

The National Academy of Sciences proposed (NAS, 1990) three principles:

1. Start with the simplest description of what is known, so that the largest and most significant uncertainties can be identified early in the program and be given priority attention;

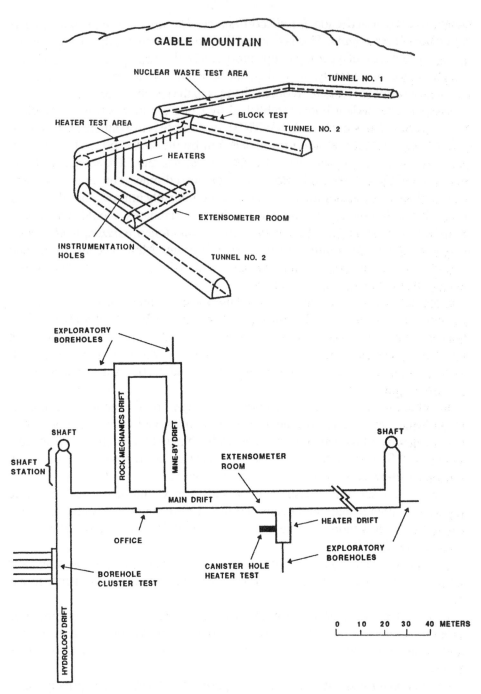

Figure 6.14. (a) Near surface test facility at the Basalt Waste Isolation Project (after Bieniawski, 1985). (b) Underground test facility concept proposed (and cancelled) for the exploratory shaft program at the Basalt Waste Isolation Project (after Bieniawski, 1985).

2. Meet problems as they emerge, instead of trying to anticipate all the complexities of the natural geologic environment; and

3. Define the goals broadly in terms of ultimate performance, rather than immediate requirements, so that increased knowledge can be incorporated in the design.

In essence, these principles ask: *Is it necessary to prescribe one engineering design, rather than allowing for alternative designs to accomplish the same goal? What can be done to accommodate design changes necessitated by surprises during construction?*

Although this sounds promising, given the history of radioactive waste management in the United States, it is doubtful that such an alternative flexible approach will be implemented. It must be remembered that since the advent of the 1982 Nuclear Waste Policy Act, nine potential sites have been identified for the first geologic repository, of which three were selected for detailed site characterization at a price tag of $1,000 million each: Hanford, Washington, in basalt; Deaf Smith, Texas, in bedded salt; and Yucca Mountain, Nevada, in welded tuff. At that stage, the deadline for acceptance of waste for disposal was January 31, 1998. Extensive site exploration and design studies were performed for the Hanford site (Bieniawski, 1985; Schmidt, 1988), known as the Basalt Waste Isolation Project (see Figure 6.14) while the other two sites also had major investigations and studies in progress or planned when the whole approach was changed in 1987 with the designation by Congress of only one site, Yucca Mountain in Nevada, for full site characterization. '*Site characterization*' in this context is a program of studies directed to collecting the geologic information necessary to demonstrate the suitability of the site for a repository, to design conceptually the repository and the waste package, and to prepare an environmental impact statement.

Site characterization at Yucca Mountain included an Exploratory Shaft Facility (ESF) featuring two drill-and-blast vertical shafts accessing the repository horizon in the most favorable welded tuff zone. Following the establishment of the Nuclear Waste Technical Review Board, this concept of shafts was abandoned in 1990 after an alternative investigation of 34 options for the ESF (renamed Exploratory *Studies Facility*). The top option now features machine excavated ramps (Stevens and Costin, 1991). Combined with the new DOE management reorganization, the deadline for the first repository in operation was moved to the year 2010. By now, about $2,500 million has been spent by DOE and its contractors since 1982, money derived by taxing nuclear power utilities at the rate of $0.001 per kWh generated. Current payments bring over $500 million per year with $5,000 million collected since 1982. The life-cycle cost of the repository at Yucca Mountain would be $7,000 million.

6.4.2 *Status*

The largest nuclear waste disposal project in the world is clearly in a state of flux.

The process of site characterization and design at the Yucca Mountain site is bedevilled by political, legal, regulatory and technical conflicts (Fairweather, 1991) – many of them attributable to the NIMBY syndrome – 'not in my backyard.'

In fact, the ten-year task of characterizing the Yucca Mountain site at a cost of $3 billion is ready to start, but cannot proceed because the state of Nevada has denied air-quality permits for dry drilling at the site. The result: the 2010 deadline could be in jeopardy. The conflicts are:

Political. The legal battle between once-receptive Nevada and the federal government dominates the situation since Nevada has refused to give DOE the necessary permits to work at Yucca Mountain.

Legal. Other state-federal issues are related to the laws which many states have banning transport of radioactive wastes within their borders (e.g. Idaho).

Procedural. There are conflicting regulations of DOE, NRC and EPA as they relate to site selection. Only in July 1990 one NRC regulation was clarified stating that the waste package provide waste containment not for 300 to 1000 years, as was understood before, but that this is the minimum performance and the DOE may take credit for waste package lifetimes exceeding 1,000 years. Many other issues are still awaiting clarification.

Technical. Doubts have been raised about the long-term geological stability of Yucca Mountain and the coupling of thermomechanical, hydrogeological, seismic and tectonic effects requires intensive research (Fairweather, 1991).

While the various debates continue, about 22,000 tons of waste sit at 60 utilities and will double by the year 2010. Additional waste from US defense facilities, also intended for the repository at Yucca Mountain, should be up to 8,800 tons by then. NRC (1990) reported that on-site storage systems are safe for at least 100 years while NWTRB (1990) pointed out that research conducted in Sweden since 1977 has suggested that containers can provide at least 10,000 years of isolation for 40-year-old spent nuclear fuel. Consequently, the containers sitting at utilities (initial temperature of up to 200°C) are undergoing a useful process of being cooled down to a 'storable' temperature of 80°C as perceived by Sweden. Clearly, the concept of monitored retrievable storage (MRS) has considerable merit for either near surface storage or controlled cooling in the geologic repository. Moreover, a more robust waste container which could isolate radioactive waste for over 10,000 years should be designed. Note that Sweden (which has only 12 reactors in operation and none planned) uses costly copper canisters placed below the groundwater level. The Yucca Mountain site is above the groundwater in an oxidizing environment unsuitable for copper.

6.4.3 *Design methodology*

The case of designing a nuclear waste repository is fraught with challenges as well as imponderables. While many nations are committed to high-level nuclear waste disposal underground, they are waiting for the USA to act first. The author has been involved in the US program for ten years and has seen concepts and deadlines come and go primarily because of the lack of clear goals and suitable methodology. With his considerable experience with the Basalt Waste Isolation Project (BWIP) during 1982-1988 and the Yucca Mountain Project (YMP) 1989-1991, he selected this case for a discussion of the appropriate functional requirements, constraints and design components in accordance with the design methodology proposed in Chapter 4. It will be clear by now that because of the political, legal and procedural conflicts, no clear technical objectives have been formalized so that it would be impossible to 'solve this case' by going through the complete design chart in Figure 4.1. Nevertheless, it should be possible to identify examples of functional requirements, constraints and design components based on perceived needs. For record purposes, design methodology has been addressed at BWIP as well as at YMP, even if in the case of the former (Schmidt, 1988) it was left incomplete due to the cancellation of the project, while for the latter it is either already out of date due to the 'winds of change' (Richardson, 1988) or still under review (Hardy et al. 1990).

Let us nevertheless attempt to formulate some functional requirements, constraints and design components based on the latest regulatory situation for a nuclear waste repository in the United States.

The purpose here is to stimulate discussion on this topic and for those in a position to do so, to initiate a comprehensive and achievable design effort which is beyond the scope of this section.

Statement of the problem. An underground storage and disposal facility is to be designed which would isolate an estimated 87,000 tonnes of high-level nuclear waste. The construction must start by the year 2004 and be ready to receive the waste by the year 2010. The overall objective is to minimize cumulative release of radionuclides to the accessible environment, such that:

1. The total system, consisting of the natural geologic system, the mined repository, and the waste package, must isolate the waste for at least 10,000 years.

2. The package is to contain the waste for at least 300 to 1,000 years.

3. The rates of radionuclide release from the engineered system are not to exceed one part in 100,000 per year for each radionuclide, after the containment period.

4. The pre-emplacement groundwater travel times from the repository to the accessible environment, over a distance of 10 km, are to exceed 1,000 years.

5. The engineered barrier system must be so designed that the waste can be retrieved for a period of 50 years after initial placement.

Examples of functional requirements. The main functional requirements that can satisfy the perceived needs are as follows (note FRs hierarchy):

FR_1: Isolate the radioactive waste from the accessible environment for 10,000 years.

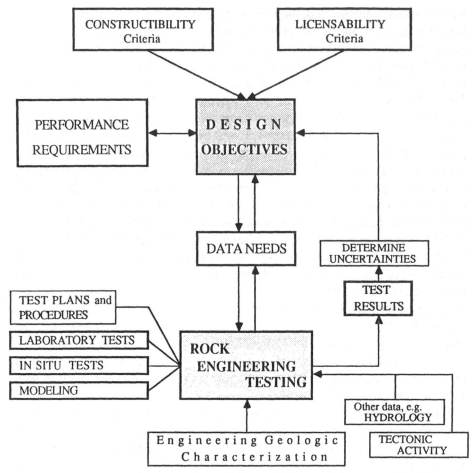

Figure 6.15. Integration of geotechnical activities with design objectives and performance requirements for repository design.

Sub-FR$_{(1.1)}$: Design a waste package to contain HLW for at least 1,000 years.

Sub-FR$_{(1.2)}$]: Ensure that the release rates for radioactive materials are not to exceed one part in 100,000 per year after the containment period.

FR$_2$: Maintain retrievability of HLW for 50 years.

Constraint 1: The associated cost must not exceed the total sum derived from taxing nuclear power at the rate of $0.001 per kWh generated (current income is $500 million per year).

Constraint 2: Site selection: groundwater travel time over 10 km must exceed 1000 years.

Collection of information. The following information is required:

1. Geologic site characterization.

2. Characteristics and quantities of waste and waste packages.

3. Repository design requirements for construction, operation, closure and decommissioning.

4. Development and demonstration of required equipment.

5. Design analyses including impact of rock mass characteristics, hydrology and tectonic activity (as per Figure 6.15).

6. Identification of technologies for surface facility construction.

7. Identification of technologies for underground facility construction, operation and closure.

8. Determination that the seals for shafts, drifts and boreholes can be emplaced with reasonably available technology.

Examples of design components. One major variable in repository design will be the behavior of rock masses under long-term loading. The second variable is the effect of high temperatures on rock masses. The third variable which the repository will experience will be radiation.

Conceptual design of the repository will resemble a room-and-pillar mine with very large flat pillars in which the waste packages will be placed horizontally. After the retrieval period, all drifts, shafts and boreholes will be sealed. The main design components are:

DC_1: Waste isolation for 10,000 years by permanent, non- corrodible container, made of stainless steel.

DC_2: Selection of the repository horizon in a rock formation above the groundwater level.

DC_3: Controlled cooling of the containers at a monitored retrieval storage area either near-surface or in the repository to allow for cooling and inspection.

DC_4: Thermal effects of HLW on drift stability during repository construction and operation to be minimized by large spacing of the container holes so that drift temperatures will not exceed 100°C.

DC_5: During the retrievability period, underground ventilation and cooling systems to keep drift temperatures at 35°C which will enable the maintenance of drift stability with readily available technology (as in deep level gold mines in Africa).

Comparison of design components with functional requirements:

DC_1 satisfies FR_1, including sub-$FR_{(1.1)}$.

DC_2 satisfies sub-$FR_{(1.2)}$: negligible groundwater flow.

DC_3, DC_4 and DC_5 satisfy FR_2.

Examples of construction procedures. Design components DC_1 and DC_2 do not require CPs but, for example, DC_4 does.

CP_1: Circular cross-section to optimize drift stability.

CP_2: Construction by a tunnel boring machine to minimize damaged rock zone.

Comparison of design components with construction procedures:

CP_1 and CP_2 satisfy DC_4. All FRs are thus satisfied.

6.5 DESIGN CASE STUDY 4: DESIGN OF A B.S. DEGREE IN DESIGN ENGINEERING

Suh (1990) pointed out that not only do people design machines, structures, processes, and software, they also design organizational systems such as industrial firms, universities and government agencies. In fact, he presented case studies involving an organizational design of a government agency (NSF Directorate of Engineering) and a college of engineering. This latter example deserves closer scrutiny.

Suh's work demonstrated that the efficiency and effectiveness of an organization depend on its design; the goals of the organization must be clearly established, as they are the functional requirements FRs of the organization in the functional domain. The design cpmponents DCs are the sub-units of the organization which are the physical units established to achieve the stated FRs. The DCs must be so chosen that the design solution represents an *uncoupled organization*, i.e. each unit knows its exact mission and can pursue its objectives. In a *coupled* organization, *all units are trying to achieve the same set of goals, thereby creating a great deal of confusion, turf fights, and inefficiency* (Suh, 1990).

The author believes that the above design approach for an organization, should apply to designing a degree program at a university, which must also have clearly established FRs to be met by DCs in the form of the courses taught. As no such baccalaureate program exists specifically in design engineering at any university known to the author, he sees it as the culmination of his efforts in improving engineering design, to present a case study devoted to designing a B.S. degree in design engineering. This would also answer a plea by Dixon (1991) who showed that the general state of engineering design education in the United States is 'intellectually stagnant' and poses a very serious problem.

But first, it is useful to review some of the reasoning used by Suh (1990) in his case studies of designing an organization, since a college department or a degree program are also organizations. He pointed out that more than 75% of engineers in the United States have terminated their formal education after receiving degrees at the B.S. level. Therefore, in engineering, the B.S. degree is *de facto* the professional degree, whereas in the pure sciences, a Ph.D. degree is required to practice in various fields. Accordingly, in the author's view, a design engineering degree must be at the B.S. level and not a graduate degree.

Secondly, organizational structures of many universities are notorious for duplication of effort and obsolescence. Due to tradition and fierce individualism, it is more difficult to reorganize universities to accommodate changing needs than any other organization. Thus, often new programs or laboratories are just added to an existing structure rather than looking at the entire organizational issue from the design point of view. To determine FRs in a university structure, Suh (1990) noted that some departments in a typical engineering college deal with engineering functions while others address products. For example, mechanical engineering is concerned with functional issues related to energy conversion and mechanisms, electrical engineering addresses issues related to electrical forces and devices, civil engineering deals with the function of solving public infrastructure problems, while chemical engineering is concerned with the production of materials and energy through chemical means. Mining engineering also addresses functions: extraction of minerals from the earth. On the other hand, nuclear engineering, aerospace engineering and marine engineering are clearly concerned with products rather than functions: their products are nuclear reactors, airplanes, and ships. Of necessity, these product departments must cover all functions of engineering since it is inconceivable to teach the design and manufacture of airplanes without including most of the functions covered in the 'function' departments. Therefore, there is an inherent conflict: product departments must duplicate some of the topics covered by function departments.

According to Suh (1990), in an engineering college, the FRs are either functions or products, the DCs are the departments, and the design equation may be written as:

$$
\begin{Bmatrix}
\text{Mechanisms} \\
\text{Chemical processes} \\
\text{Electromagnetism} \\
\text{Nuclear power plant} \\
\text{Shipbuilding} \\
\text{Airplane manufacture}
\end{Bmatrix}
=
\begin{bmatrix}
\times & 0 & 0 & \dots & \times & \times & \times \\
0 & \times & 0 & \dots & \times & \times & \times \\
0 & 0 & \times & \dots & \times & \times & \times \\
\times & \times & \times & \dots & \times & 0 & 0 \\
\times & 0 & \times & \dots & 0 & \times & 0 \\
\times & 0 & \times & \dots & 0 & 0 & \times
\end{bmatrix}
$$

$$
\times
\begin{Bmatrix}
\text{Mechanical Engineering Department} \\
\text{Chemical Engineering Department} \\
\text{Electrical Engineering Department} \\
\text{Nuclear Engineering Department} \\
\text{Naval Architecture and Marine Engineering} \\
\text{Aeronautical Engineering Department}
\end{Bmatrix}
$$

To understand the design equation, it may be recalled that design was defined by Suh (1990) as a mapping process between the functional requirements (FRs) in the functional domain (design objectives) and the design components (DCs) in the physical domain (design solution). This relationship may be characterized mathematically. Since the characteristics of the required design are represented by a set of independent FRs, these may be treated as a vector FR with m components. Similarly, the DCs in the physical domain also constitute a vector DC with n components. The design process then involves choosing the right set of DCs to satisfy the given FRs, which may be expressed as

$$\{FR\} = [A]\{DC\}$$

where [A] is the design matrix. When $m = n$, [A] is a square matrix and the above relationship is called the design equation. For example, when $m = n = 3$, [A] may be written as

$$[A] = \begin{bmatrix} A_{11} \ A_{12} \ A_{13} \\ A_{21} \ A_{22} \ A_{23} \\ A_{31} \ A_{32} \ A_{33} \end{bmatrix}$$

The left-hand side of the design equation represents *what we want in terms of design goals*, and the right-hand side of the equation represents *how we hope to satisfy the FRs*.

The simplest case of design occurs when all the nondiagonal elements are zero; then the design equation may be written for the case $m = n = 3$ as

$$FR_1 = A_{11} \, DC_1$$
$$FR_2 = A_{22} \, DC_2$$
$$FR_3 = A_{33} \, DC_3$$

This design satisfies the Design Principle 1 because the independence of FRs is assured when each DC is changed. This situation results in an *uncoupled* design. In a *coupled* design, the design matrix consists of mostly nonzero elements. In this case, for example,

$$FR_1 = A_{11}DC_1 + A_{12}DC_3 + A_{13}DC_3$$
$$FR_2 = A_{21}DC_1 + A_{22}DC_2 + A_{23}DC_3$$
etc.

and a change in FR_1 cannot be accomplished by simply changing DC_1 since this will also affect FR_2.

We can now go back now to the design equation for the engineering college. It is obvious that the design matrix listed by Suh (1990) is highly coupled and is not a diagonal matrix. To design an uncoupled organization for an engineering school, one would require the freedom to define the FRs and the constraints. Such freedom seldom exists within a university setting but much greater latitude is available when it comes to structuring a particular degree program. Accordingly, we may realistically design a B.S. degree program in design engineering.

The term *design engineering* is selected to emphasize its importance on a par with any major branch of engineering represented by either a function or a product department such as civil or nuclear engineering, respectively. By analogy, the design engineering department of the future will be as separate as an engineering science department is today. The goals of the design engineering program will be the FRs while the DCs – representing the design solution – will be the specific courses offered to satisfy the FRs.

In considering the goals of a B.S. degree in design engineering, it is useful to recall that Dixon (1991) pointed out that 'for nearly 40 years, it has been generally agreed that a proper goal of undergraduate education is to provide students with an understanding of the fundamentals of engineering science.' As a result, the teaching of engineering science continues to be a tremendous success and well-written and time-tested textbooks are readily available.

On the other hand, engineering design education is in a state of 'intellectual- and hence curricular-stagnation' (Dixon, 1991) because the design education community, in stark contrast with what has occurred in engineering science, never developed a consensus on what constitutes the fundamentals of the field of engineering design. Not surprising, too, that textbooks on design engineering are lacking.

However, as listed in Table 5.8, many design fundamentals have now been identified and they can form the backbone for the goals of our B.S. degree program in design engineering. The guiding objective of the program will be not actual design *experience* but the explicit learning of design concepts and principles. Teaching current practice will be a second priority. Accordingly, our FRs will be as follows:

FR_1 = proficiency in engineering science fundamentals.

FR_2 = social and economic implications of design.

FR_3 = design theory and methodology.

FR_4 = product realization processes in business organizations, including project management.

FR_5 = manufacturing and construction processes.

FR_6 = design of components, including computer aided design.

FR_7 = multi-functional team design.

FR_8 = case studies of best design practice.

FR_9 = information retrieval and learning.

FR_{10} = principles of creativity and innovative design.

Note that these FRs actually define what design engineering is: it is *not* an art or a skill that can be learned only through experience; it *is* a cognitive (intellectual) activity based on knowledge: of facts, science, engineering science, design, manufacturing and construction principles, processes, strategies and business organizations (Dixon, 1991).

The constraints for the B.S. degree in design engineering are:

C_1 = admission to the junior year of the design engineering major restricted to students with a minimum GPA = 3.0.

C_2 = cumulative GPA = 3.0 required for graduation.

C_3 = presentation of a senior design report to a design jury consisting of practicing design engineers.

The design solution is characterized by ten DCs as follows:

DC_1 = acceptance of students qualifying for engineering science and engineering majors honors programs.

DC_2 = fulfillment of general education requirements in arts, social studies and humanities, including history of technology.

DC_3 = a 3-credit course on design principles and methodologies in engineering and architecture.

DC_4 through DC_9 = six specific 3-credit courses one for each FR, from FR_4 through FR_9 in accordance with Table 5.9.

DC_{10} = a 3-credit course on principles of creativity and innovation: idea generation, communication and management of design innovation.

An important aspect to realize is that the purpose here is not to make ready-to-work designers of students in four years. The role of the university must be to teach the fundamental knowledge, strategies and principles on which design is based, subsequently industry must do its share concerning design practice.

The design equation for our BS degree in design engineering is a diagonal matrix depicting uncoupled design:

$$
\begin{Bmatrix} FR_1 \\ FR_2 \\ FR_3 \\ FR_4 \\ FR_5 \\ FR_6 \\ FR_7 \\ FR_8 \\ FR_9 \\ FR_{10} \end{Bmatrix} = \begin{bmatrix} \times & 0 & 0 & 0 & 0 & 0 & 0 & 0 & 0 & 0 \\ 0 & \times & 0 & 0 & 0 & 0 & 0 & 0 & 0 & 0 \\ 0 & 0 & \times & 0 & 0 & 0 & 0 & 0 & 0 & 0 \\ 0 & 0 & 0 & \times & 0 & 0 & 0 & 0 & 0 & 0 \\ 0 & 0 & 0 & 0 & \times & 0 & 0 & 0 & 0 & 0 \\ 0 & 0 & 0 & 0 & 0 & \times & 0 & 0 & 0 & 0 \\ 0 & 0 & 0 & 0 & 0 & 0 & \times & 0 & 0 & 0 \\ 0 & 0 & 0 & 0 & 0 & 0 & 0 & \times & 0 & 0 \\ 0 & 0 & 0 & 0 & 0 & 0 & 0 & 0 & \times & 0 \\ 0 & 0 & 0 & 0 & 0 & 0 & 0 & 0 & 0 & \times \end{bmatrix} \times \begin{Bmatrix} DC_1 \\ DC_2 \\ DC_3 \\ DC_4 \\ DC_5 \\ DC_6 \\ DC_7 \\ DC_8 \\ DC_9 \\ DC_{10} \end{Bmatrix}
$$

For clarity, Table 6.14 gives more details of the specific curriculum requirements. Note that this table differs from Table 5.10 which was directed at a professional 5-year degree of the distant future, culminating in an M.S. degree.

Some further comments are warranted. Crochetiere (1987) described engineering design education at Tufts University which offers M.S. and Ph.D. degrees in engineering design. Although no undergraduate degree in design engineering exists, there is a separate Department of Engineering Design at Tufts. Design related courses, offered to students enrolled in the accredited *undergraduate* engineering programs, have these features:

1. The open-ended type of design problem for which there is no single solution;

2. The active involvement of the student researching a problem area and deciding upon the form of the solution; and

3. Presentation of design solutions to a design jury made up of practicing engineers.

The significance of the design programs at Tufts is that the creation of the Department of Engineering Design has not resulted in 'turf battles' but – on the contrary – has stimulated the development of a number of related research activities and courses in other engineering as well as non-engineering departments (e.g. a joint program in 'human factors engineering' with the liberal arts college). Accordingly, it could be expected that a B.S. degree in design engineer-

Table 6.14. Course listing for BS degree in design engineering.

Total credits: 128		Total courses: 40	Length of study: 4 years
Mathematics	16		
Basic sciences	23	(chemistry 8, physics 12, economics 3)	
Engineering sciences	37	(mechanics 8, electro/magnetism 7, thermo/fluid/heat 10, materials 6, computer science 6)	
Engineering design	27	(seven 3-credit courses: design theory, plus six courses in Table 5.9; eng. graphics 3, and project management 3)	
Senior design project	7		
Arts/Humanities/Social studies	12	(includes two writing-intensive courses, one course in history of technology and design, and one course in creativity and innovation)	
Foreign language	6		

Note: Course numbers are arbitrary but, for identification purposes, they conform to course descriptions at Penn State University.

1st Semester		2nd Semester		3rd Semester	
Math 140	4	Math 141	4	Math 230	4
Chem 12	3	Chem 13	3	Phys 204	4
Chem 14 lab	1	Chem 15 lab	1	E. Mech 110	5
Eng Graphics	3	Phys 202	4	Eng. Economics	3
Foreign language	3	Foreign language	3	Total	16
Comp. Sci. 201	3	Total	15		
Total	17				

4th Semester		5th Semester		6th Semester	
Math 251	4	Aero 308 (Fluid)	3	Mat Sci 416	3
Phys 237	4	ME 120 (Thermo)	4	EE 340 (El/Magn)	3
E Mech 112	3	EE 203 (El/Magn)	4	Design 202	3
Comp Sci 407	3	Mat Sci 414	3	Design 303	3
A/H/S	3	Design 101	3	A/H/S	3
Total	17	Total	17	Total	15

7th Semester		8th Semester	
Design Project	3	Design Project	4
Nuc E 307 (Heat)	3	Design 406	3
Design 404	3	Design 407	3
Design 405	3	Project Management	3
A/H/S	3	A/H/S	3
Total	15	Total	16

ing, as designed by the author, would readily fit into any university or engineering college structure provided that the academic 'movers and shakers' have sufficient vision and determination to initiate the necessary action.

Eder (1990) presented an argument in favor of a comprehensive design curriculum serving as a core for any specialized engineering program; since many engineering subjects are duplicated in separate courses (usually without cross-referencing between them), an engineering design core could be a catalyst between engineering and design. He did not, however, give a specific example showing just how such a curriculum would fit into an engineering major but listed ten design-oriented courses in two groups. The basic group consisted of these topics: introduction to engineering design, introduction to theory of technical systems, introduction to modeling and computation, exercises in design theory and practice, special design knowledge, and special families of technical systems. The 'main' group of topics consisted of: advanced design science, advanced design knowledge, design of special technical systems, and advanced representation and CAD. No details of the content of these topics/courses were given by Eder (1990) but a clear message has emerged which supports the author's contention that the time is ripe for introducing a B.S. degree in design engineering.

In conclusion, the proposed B.S. degree in design engineering will achieve what Dixon (1991) so effectively stipulated: 'To get started in a design role in industry, students need to know that the knowledge they have is correct and useful, that it is okay not to know everything, and that they have the background and ability to learn new facts, methods, processes, and principles as needed.' For those more ambitious, the B.S. degree in design engineering will be an excellent stepping stone for an M.S. or even a Ph.D. degree in this field – two programs already in existence, e.g. at Tufts University.

It is hoped that this case study will stimulate discussion among engineering professors as it is emphasized that this proposal for a B.S. degree in design engineering is open to suggestions and constructive criticism.

CHAPTER 7

Skills development

If money is the only hope for independence, you will never have it. The only real security that a person can have in this world is a reserve of knowledge, experience and ability.

Henry Ford

Having discussed design theory, education and practice, it should be readily apparent that our search for innovative designs will be much enhanced by developing and improving our creative skills as designers. It is pointed out in Chapter 2 that creativity and innovation are derived from three main sources: inherent ability, motivation, and acquired skills. Clearly, while only a select few may be born creative geniuses, the door to innovation in design is open to all others who are motivated and prepared to develop their skills.

Therefore, a discussion of the various aspects of skills development is not only appropriate but will also be cost-effective for the reader since the generous supply of commercial courses and materials on this topic is not matched by their low cost of attendance!

The author has conducted a series of practical workshops in skills development across America and abroad, and has found that three topics deserve particular attention as related to design: generation of ideas, communication of ideas, and management of design innovation. They are presented in this chapter in that order, followed by examples from actual *practicum* sessions at the author's workshops. The material is presented in a form most suitable for assisting a reader subscribing to the principle of 'learning by doing.'

7.1 GENERATION OF IDEAS

'Good ideas are a scarce commodity, arising only in the right institutional climate', commented Barr (1988) when discussing the process of innovation in American corporations. 'But the human brain won't create automatically; one must want to create and work at it,' wrote Montalbo (1988). These statements are compatible with what is emphasized in Chapter 2 that creativity requires practice, persistence and patience, and that – unfortunately – objections to new and creative ideas tend to be common in virtually any workplace. Yet top corporate leaders at leading companies believe that *creativity is not optional; it is a survival skill.* Someone even said: 'Necessity may be the mother of invention but imagination is the father!'

Rather than overwhelm the reader with a multitude of techniques for idea generation, let us consider the approaches that are most effective and constitute a good start; once more proficient and 'hooked,' the reader may expand his or her knowledge by consulting the references provided in this chapter. Remember what Einstein once said: 'Curiosity, obsession and dogged endurance, combined with self-critique, brought me my ideas!'

Before concentrating on specific group techniques, such as brainstorming, there are a number of common-sense stimulants to generate ideas on an individual basis that are worth identifying. These are (Montalbo, 1988):

1. *Ask yourself 'why not?'* By meditating over a series of 'why not?' questions and answering them, a new idea may emerge. For example, 'why not change the form?' led to the marketing of usually granulated sugar in a powdered form and also compacted into cubes – both highly successful forms of the same product.

2. *Try 'supposing.'* Edison, who created more than a thousand inventions, asked himself 'suppose that' or 'what if' and tried various ideas in building many hypotheses. While this may also result in unworkable alternatives, it may trigger useful ideas that might otherwise have been overlooked.

3. *Experiment with 'word magic' and 'visual thinking.' Open a dictionary at any place, choose a word and start making associations, a new idea may be born out of an unexpected word. Visual thinking is the use of freehand drawing which often stimulates ideas because it forces the focusing of the mind on specific tasks.*

4. *Try the 'reversal' principle.* An inventor saw the wind blow dust around; he reversed that action and created the vacuum cleaner to pick up dust. It was known long before Faraday that an electric current could magnetize pieces of iron or steel placed near it. Scientist Michael Faraday reversed the process: he passed wire through a magnetic field and produced electricity.

5. *Apply the 'adaptation' principle.* From the ball-point pen which rolls ink on paper, to the deodorant which rolls liquid on the body, to the lipstick which rolls color on the lips, all three ideas are adapted from one basic concept: the roll-on idea. Another example is Henry Ford's mass production concept; he adapted the conveyor belt and assembly line to automobile production from the meat-packing industry. The adaptation principle also means building upon one's own experience, and on the experience of others.

6. *Try the association technique.* Linking an earlier idea with another is a fast and easy way to develop additional ideas; from one idea or product one can create additional ideas by building upon previous ideas. This concept is widely used in group brainstorming.

7. *Exaggerate.* Advertising people and comedians know the value of exaggeration in creating humor. Exaggeration can also lead to new ideas.

8. *Use analogy.* By using an analogy, one can reason from what is known to

what is unknown. For example, the invention of the stethoscope for listening to sounds within the body is credited to a physician who observed two boys at a school yard using a seesaw: one boy held his ear to one end of the seesaw while the other tapped the opposite end with a rock. The analogy approach – the use of metaphors – was developed into a technique called 'synectics' by William J. Gordon and is discussed in Chapter 2. There is a story about how analogy worked for the Coors Company. This brewer was paying to dispose of gallons and gallons of spoiled – and thus worthless – beer. A creativity consultant directed the managers to a scene from 'Tom Sawyer.' Inspired by the way Tom persuaded his friends that painting a fence was a privilege they should pay for, Coors is now selling its spoiled beer to the Japanese to be used as feed for their beef cattle!

9. *Combine things*. By combining two or more things one can create new ideas: putting adhesive tape and bandages in strip form led to the Band-Aid; joining an alarm clock and a radio resulted in a clock-radio; adding an eraser to the end of a pencil provided a new product, much in demand. This is also called 'synergy,' combining existing elements in a new way. Sister Tabatha Babbett is credited with the invention of the circular blade. After watching two men laboriously sawing wood as she worked at her spinning wheel, she figured it would be easier if saw teeth were cut into the edge of a wheel.

10. *Do something different*. Merrill Lynch brokerage house used a stampede of bulls as an attention-getting device to emphasize how really bullish they were about the stock market. The concept of lateral thinking of De Bono (1967) is an example of a radically different approach for generating new ideas. Another way to be different is to fantasize. Engineers at NASA used this method to come up with 'Velcro,' the substitute for zippers and buttons, when they were trying to find a fastening device for space suits that astronauts could manipulate with their bulky gloves. One member described a fantasy of running through a forest and having thorns stick to his clothes. That led to a fastener that gripped with thousands of thornlike fibers.

Armed with these individual techniques, we are now well equipped to consider some group-effective methods for idea generation.

7.1.1 *Brainstorming*

The most common creative-thinking technique, and the oldest, is brainstorming. The term was coined by Alex Osborn, an advertising executive, in 1953 when he described brainstorming as a group activity involving these four principles: (1) do not judge ideas; (2) let your mind wander; (3) aim for quantity; and (4) hitch-hike on previous ideas.

An important prelude to creative problem-solving is to make sure that everybody understands the problem thoroughly. Sometimes problems elude solution because they are not clearly defined. The next step is to ask the group to relax, as stress is the surest way to extinguish the creative spark. Remembering that there is no right or wrong answer, a brainstorming session must not pass judgment on any ideas until many alternatives have been generated.

Hundreds of businesses and organizations use brainstorming every day to unlock creativity and to tackle tough problems. In fact, many large corporations such as Eastman Kodak, RCA and Exxon have been brainstorming for over 25 years! So, this technique is not another trendy gimmick. It is a powerful technique for problem-solving because it taps one's imagination and ideas buried in the subconscious mind.

Here are the benefits, rules and tips for group brainstorming, together with some examples of what can be achieved.

Benefits of brainstorming:

1. Increases the number of ideas; a 30-minute brainstorming session can produce as many as 100 new ideas.

2. Quality of ideas increases; as the quantity of ideas increases, useful solutions will also increase.

3. Future problem-solving is easier: often, more useful ideas are generated than are needed and an inventory of ideas will save time with future problem-solving.

4. Sense of teamwork increases.

5. New ideas trigger other new ideas.

6. Creative abilities are contagious – the more brainstorming is done, the more easily new ideas emerge as experience is accumulated.

7. Resistance to change is overcome – as momentum is gained for making changes, change becomes easier to accept because people feel that their input matters.

8. Flexibility toward problem – solving: one learns that there is more than one answer or more than one way to the answer.

Rules for brainstorming:

1. Criticism is out. Judgment is suspended until subsequent evaluation.

2. Free-wheeling is welcomed. This means the wilder the ideas the better since it is easier to tame down than to pop out ideas. Even 'off beat,' impractical suggestions by an individual may 'trigger' in other group members practical ideas which might not otherwise occur to them.

3. Quantity is wanted. The greater the number of ideas, the more the likelihood of good ones.

4. Combination and improvement are sought. In addition to contributing ideas of their own, group numbers should indicate how suggestions by others could be turned into better ideas or how two or more ideas could be combined into a still better idea.

Tips for brainstorming:

1. Be specific in problem formulation.

2. Set a time limit.

3. Include key staff people, those with good imaginations who work closely on the problem.

4. For a fresh perspective, include new people (not typically involved in the problem area).

5. Use cues to trigger thought – pictures, news clippings, music or even a visit to a problem site will stimulate thinking on the subject.

6. Hold sessions at times of day that are most convenient and conducive to creativity (avoid the 'drowsy' period right after lunch).

7. Keep it a team effort – never let any individual, even a session leader, dominate. Everyone should be free to give as much as they can.

8. Document the session.

9. After brainstorming is over evaluate and prioritize the ideas.

10. Make it a habit to brainstorm on your own: be prepared to catch fleeting ideas any time, any place. Carry a note pad or a pocket tape recorder to get your ideas down while they are fresh in your mind.

Examples of ideas from brainstorming
The amazing thing about brainstorming is how many ideas can be generated by a few people in just one session. Moreover, most people do not consider themselves particularly creative: on a scale of 10, most participants at the author's innovative design workshops declared themselves to fall in the range 4-7; only a few had the courage to designate themselves as an 8, but none above that! An excellent starter to jog people's imagination is to ask a question like *What is a half of eight?* and instruct them to be creative. The poem below shows the possibilities.

> '½ of 8'
> *What is one half of eight?*
> *Four – I guess.*
> *No more? No less?*
> *But wait!!!*
> *½ of 8 is 3 or E.*
> *Tipped over, it's m or w – see?*
> *To the Romans it still was VIII.*
>
> *It's a bird in flight, the world,*
> *the sun, a full moon at night; a*
> *pair of sky divers holding hands.*
> *It's a ball, a balloon, a raindrop,*
> *a teardrop, a drop of dew; it's the*
> *yolks of two eggs staring at you.*
> *It's a plate or a saucer, even a ring;*
> *half of eight is a tricky thing!!!*
>
> *But after I've pondered, thought of it all,*
> *½ of 8 was nothing at all – zero!*
>
> M.G. Nickoll (San Diego)

As a start to brainstorming, a 'warming-up' session aimed at acquainting the participants with the process of brainstorming might be devoted to finding a way to improve a 'coffee cup for the 21st century.' In a 10-minute session each participant was instructed to generate a minimum of 10 ideas but many doubled it by coming up with such 'winners' as: cups that are self-cleaning, self-filling, coffee-strength regulating, and beeper-equipped for locating!

As a next step to consolidate the initial experience and put the group into a relaxed even playful or humorous mood which would encourage idea generation, the brainstorming session is split into two or three groups of 4-6 participants each. The groups then brainstorm the same question: 'What should the ideal bathroom be like?' and compete for the best three ideas from each group. At the author's workshops, a 20-minute session with such a setting might produce between 30 to 50 ideas per group! Some were truly innovative and are quoted here to demonstrate the spirit of human creativity striving to be liberated! For example, an ideal bathroom should have holographic scenery, a coffee-dispenser, be self-cleaning, involve no water but rely on chemical processes, should monitor and analyze the content of body liquids, be equipped with a library of page-turning books, have a button to start a car remotely (to warm it up in winter), and should even sound a welcoming trumpet upon entering it!

Such a relaxing exercise invariably generates a conducive atmosphere for serious brainstorming of a pertinent question. For example, sessions at the design workshops for the Bureau of Mines staff were directed at answering two questions: 'What should be the real long-term goals of the Bureau of Mines beyond 5 years?' and 'What are the best ways to achieve the most effective technology transfer from research to practice?' Typically, the first question generated between 80 to 100 ideas from 12 people in 45-minutes. After evaluation and prioritizing, the top ten ideas were compiled as depicted in Table 7.1. Similarly, Table 7.2 presents the prioritized results of brainstorming the second question which generated between 60 and 80 suggestions from the same group of people. These same questions were put to the participants at three different Bureau of Mines research centers and the volume of creative ideas was high at all localities; there is no doubt that brainstorming is highly productive.

Nevertheless, care must be exercised in organizing brainstorming sessions. They should be well planned and feature well-defined problems. This is important because brainstorming is wide open to interpretation and experimentation, and thus allows one to develop a personal style resulting in ever better sessions. The most effective way to learn about brainstorming is to experience it.

7.1.2 *Storyboarding and the Delphi Technique*

Barr (1988) reports that a sophisticated form of brainstorming, called storyboarding, is credited to Walt Disney. Apparently before computer animation, Disney was faced with the problem of how to organize the thousands of drawings that

Table 7.1. Results from a brainstorming session on: 'What should be the long-term goals of the Bureau of Mines?' (Top ideas from 106 items brainstormed by 11 people in 45 minutes).

1. Elimination of mining hazards.
2. Get the miners out of the mines by way of: robotics, remote control, etc.
3. Most productive mines in the world.
4. World leader in mining research.
5. Alternative sources for strategic materials.
6. Increase significantly the number of researchers with a Ph.D in mining.
7. Develop innovative mining technology to tap new reserves: super longwalls, hydraulic mining, in situ extraction, undersea mining, etc.
8. New uses for US minerals and coal.
9. Recycling of mineral products for environmental protection.
10. Improve technology transfer.
11. Diversification and expansion.
12. Reduce paperwork.

Table 7.2. Results from a brainstorming session on: 'What are the best ways to achieve the most effective technology transfer from research into practice?' (Top ideas from 84 items brainstormed by 12 people in 45 minutes).

1. Directory of Bureau experts on call for mining companies.
2. Direct industry input into projects and proposals review.
3. New publication category for practical applications.
4. Toll free hotline for publications and electronic acquisition for the requester.
5. Expedite the publication process.
6. Sabbaticals or personnel exchange programs with industry.
7. Appoint technology transfer officers.
8. Improve handbooks and training programs.
9. Open house and trade-show type displays.
10. Quality promotion plan for technology transfer videos and books.
11. Subsidization of demonstration trials for cooperating companies.
12. Opportunity for foreign travel to observe mining technology abroad.

made up his cartoon features. Remembering that Leonardo de Vinci was known to have affixed his drawings to the wall in order to study them, Disney came up with the idea of a cork-covered 'infinite wall' on which his artists and art directors could pin their drawings in sequence. As ideas were generated by members of the group, the storyline of a movie would unfold before them.

At a modern storyboarding session in response to a well-defined problem, participants write their ideas down on cards or slips as fast as they can, and the cards and slips are tacked to the wall under the appropriate category. This allows the ideas to be moved around and relationships formed. After an hour or two, the process shifts from idea generation to criticism and evaluation. The Exxon Innovation Center in Houston and the Air Force's Office of Innovation in San Antonio are renowned for their use of 'blue slips' for storyboarding.

The Delphi Technique is another version of brainstorming during which the participants – usually experts in the problem area – are separated from one another and asked to generate ideas. This approach was taken because it has been shown that, although brainstorming is usually considered to be a group technique, the processes are as effective when done by a single individual. In a Delphi session, once the ideas are collected from individuals, they are distributed to all participants for evaluation.

The Delphi method can also be used by mail. Questionnaires are sent to respondents to generate ideas; the responses are summarized by a staff group. Subsequently, a second questionnaire is sent to the same participants who independently evaluate the earlier responses and vote on priorities. The staff team makes a new summary of the evaluations and a decision-making group decides on the most suitable ideas. This approach is particularly useful when highly-qualified people are unable to meet.

7.2 COMMUNICATION OF IDEAS

'Conceiving a new idea is like taking the first step on a 1,000-mile journey. The second step is figuring out how to convey it', writes Barr (1988). The ability to communicate effectively is an indispensable asset in every business. After all, even the best idea may be lost unless it is communicated with skill. Such a skill is the result of understanding, preparation and experience. Successful communication is a balanced combination of clear, sequential ideas expressed by the spoken or written word as well as in behavior, physical expression or various forms of symbolism. Communication is not just talking.

The author has subscribed to this doctrine during the past 20 years, ever since he became a member of Toastmasters International, the best organization to help people advance their communication and leadership abilities. What follows in this section are his personal guidelines and tips collected from over 150 speech-making assignments and an equal number of publications and reports.

Effective communication in our design endeavors is absolutely vital for several reasons:

1. Effective communication leads to high morale and constitutes in itself a powerful incentive at all levels.

2. Only by communication can attitudes be changed.

3. Communication is a leader's primary tool.

4. Successful communication increases understanding and trust.

5. Without planned communication, no team of people can function well.

It will be recalled from Chapter 4 that, with apologies to Albert Einstein, the effectiveness of good communication is best demonstrated by this equation:

$$E = MC^2$$

where E equals effectiveness, M is the mastery of the technical subject matter, and

C equals communication skills. Since C is squared, this factor can outweigh the importance of M!

The most important aspect of communication in an organization is that it must be a three-way process: it should pass sideways or laterally throughout the organization, downwards from one level to another, and also upwards. Interestingly enough, it is usually the upward communication which is most difficult to achieve. What is generally needed in this respect is an aptitude for asking the right questions and a willingness not just to listen, but to listen well. Moreover, for best results, personal communication – an open discussion face to face – is much better than issuing impersonal directives.

In communicating, either orally or in writing, three things come to mind out of the myriad of items influencing the process of conveying a message. These are the aspects most frequently neglected when faced with an oral presentation or a written report: understanding the audience, organization of the material, and preparation.

7.2.1 *Audience analysis*

Knowing one's audience is crucial because it determines the length of the message as well as its style and complexity. When writing or making a presentation to a board of directors on a new design project, the design engineer may be restricted by the amount of time allotted and by the level of technical expertise of the listeners. This will certainly require a different approach from the one to be used when addressing a group of one's peers. Accordingly, it is essential to take the trouble to find out what the background of the audience is and how long the presentation or report is expected to be.

7.2.2 *Organization*

Although so obvious in importance, the organization of a speech or report is often deficient with a resultant loss of clarity and hence of the opportunity to achieve the desired results. Organization of one's ideas is a difficult aspect of communication.

The essence of good organization is the creation of order out of chaos. This is done by planning three aspects: the introduction, the main body, and the conclusions. The introduction must be directed to (1) capture attention; (2) arouse interest; and (3) suggest the theme of the communication. In this respect, the very first sentence is particularly important since this is the opening. There are some commonly recognized approaches to good openings, such as a startling question or statement, an exhibit when making a presentation or an appropriate quotation or illustration. Bad practice in an opening is an apology, a commonplace statement in a commonplace manner, a story by way of illustration which does not connect, or a slow-moving sentence.

The body of the message comprises the factual material. It should give the arguments supported by illustrations and examples. Generally, it consists of three items: (1) a statement of the facts; (2) the proof for these facts; and (3) the refutation of contrary views. Another approach might be the 'Past-Present-Future' plan which would be suitable when dealing with a historical subject or changes in human affairs.

The conclusion is the climax, the clincher which gets the desired action. It must tie in with the introduction and must never leave the listener or the reader in doubt as to what is recommended. A weak, inconclusive or abrupt ending can kill the best speech or the best report. Some good types of conclusions are: (1) an appeal for action; (2) a summary of the points made, with deductions; or (3) an appropriate story, quotation or illustration.

Here are some additional tips compiled by the author as highly effective for communicating ideas in an organized fashion:

1. Be purposeful: keep the goal always in mind.

2. Be definite: generalizations do not win support from thinking people.

3. Be simple: plain words and short, direct sentences are more easily understood than long, complex ones.

4. Be fair: do not be rough with the opposing views as there may be truth on both sides.

5. Be natural: using one's best style is the best way; imitating someone else usually misfires.

Finally, a time-proven formula for effective communication is this: *state what you are going to say, say it, and summarize what you have said!*

7.2.3 *Preparation*

In a world where time is considered the most precious commodity, it is not surprising that insufficient time is spent on preparation when faced with the task of conveying an important message either orally or in writing. Yet a failure to be well prepared can have embarrassing even disastrous consequences. At professional meetings, it is painful to see speakers who are poorly organized and clearly unprepared. Worse still, they sometimes even admit that they have put the material together at the last minute! This is an offense to the audience!

The author's motto over the years has been: 'for one minute of presentation, use one hour of preparation.' For those horrified by this statement or simply incredulous, it needs to be pointed out that with a little planning and self-discipline it is possible to find the time to meet such a seemingly tough criterion! And the rewards of effective communication are really worth it!

Here are some tips for preparation:

1. Start early on planning the assignment: think about it while travelling to work or when relaxing; talk to your friends and colleagues, if appropriate.

2. Consider the type of audience.

3. Decide the length of the presentation.

4. Decide on the purpose. Exactly what it is that needs to be accomplished.

5. Write down the purpose clearly and simply. This really is the conclusion of the presentation. The *first* thing to plan is the *last* thing one intends to say.

6. Make a list of all you know about the topic.

7. Plan to introduce the subject so as to gain attention.

8. Prepare a plan to lead from the envisioned opening to the already decided upon conclusion; this will constitute an outline.

9. Go through the outline repeatedly and with care to test whether it is logical, interesting and convincing.

10. Write the speech or report in full.

11. If this is a speech, prepare simple notes featuring the headings only and practice the speech in front of a mirror. Time it to check if it is within the time limit.

12. If this is a report, be prepared to work over several drafts for clarity and to stay within the expected length.

Mastering audience analysis, organization of the material, and preparation will ensure the effective communication of creative ideas and will go a long way towards the realization of an innovative design. However, even the best of design projects can be jeopardized if mismanaged, and hence acquiring the appropriate skills for managing design innovation is an essential undertaking. This means, for both leaders and followers, understanding what constitutes good management.

7.3 MANAGEMENT OF DESIGN INNOVATION

It is not intended in this section to deal with the whole subject of management to which prominent business schools devote many years. Having managed research and design projects for 22 years, the author is convinced that a few basic guidelines – for those in leadership positions as well as those practicing design engineering – will be appropriate. The specific topic of managing design innovation has only recently acquired the attention it deserves. In fact, some leading engineering universities in the United Kingdom, such as the University of Strathclyde, offer special professional courses and workshops, entitled: 'Total Design for Managers' (Pugh, 1989). The London Business School offers this advice to its graduates: 'Design is a central activity to line managers and it contributes directly to company profitability' (Haley, 1989). A Harvard Business Review classic (Wrapp, 1984) gave this advice: 'At executive levels, good managers don't make policy decisions – rather they give a sense of direction and are masters at developing opportunities.'

This last statement is particularly fitting to the management of design innovation and it leads to this question: 'Why must we manage design projects more effectively?' Randolph and Posner (1988) provide the answer by pointing out the rapid technological changes that we continue to experience; every year one of every eight jobs in the USA did not exist before and one of every nine jobs is eliminated. A recent survey revealed that fewer than 25% of employees say they are working near full potential; half of those surveyed do only what is required to keep their jobs and 75% said they could be significantly more effective.

In essence, two major issues are critical to good management of design innovation: (1) leadership skills on the part of the managers, and (2) understanding and supporting good management initiatives on the part of those being managed.

7.3.1 *Leadership skills*

We hear so much about leadership today and in so many areas – politics, business, academia, personal affairs. Applied to professional and daily life, leadership may be viewed as (Clark and Clark, 1990):

1. The ability to obtain the maximum support and effort from a group;

2. The creative ability to see a problem, recognize it, plan a solution, and execute that solution without having to be prompted by someone else;

3. Leadership is the lifting of a person's vision to higher sights, and the raising of a person's performance to a higher standard;

4. Leadership is the ability to make people want to do things they would not ordinarily think of doing, and the ability to make people want to accept as their own, the objectives of the enterprise of which they are a part.

But how does one acquire these skills? There are many views on this subject but those of Fiedler (1980) are particularly worth considering. He points out that some people are leaders because they exercise power but leadership and power do not *have* the same meaning: leadership involves inducing others to follow when they do not have to. On the other hand, a situation may also create the leader but a leader in one situation may not be a good leader in a different situation regardless of the traits he or she possesses. Apart from the power and situation aspects, there are four other schools of thought on leadership, namely, that leaders are born not made; that leadership is teachable to all who want to learn; that the really critical ingredient of leadership is vision; and that leadership intrinsically involves ethical reflection. Each of these six propositions makes an important contribution but it seems that standing alone each is inadequate and all six should be viewed as supplementing one another.

Nevertheless, the best advice comes from Donald M. Kendall, the retired chairman of Pepsi Cola, Inc., who had this to say when asked what it takes to become a successful corporate leader: 'There is no place where success comes before work, except in the dictionary. You can't get to the top of any profession without a lot of hard work, and I don't care if you are in art, music, in business, or in the academic world. It also requires enthusiasm and excitement about what you are doing. If you are not happy every morning about leaving for work, you're not going to be successful. The other thing you need, in probably equal proportion, is a lot of luck because there are a lot of people with the same ability, who work very hard, who haven't made it, and who deserve to make it.'

With this background in mind, when it comes to managing design innovation, one may wonder what is the difference between a manager and a leader. Fiedler (1980) defined a leader as someone who directs people and pointed out that there are many managers who are not leaders because only that part of a manager's job that concerns supervision of people is leadership. Making phone calls, going to meetings, writing reports, looking after budgets may be a manager's tasks but they do not represent leadership. On the other hand, there may be leaders who do not have a manager's function, e.g. being the captain of a football team. Accordingly, since management involves the customary five functions: planning, organizing, staffing, directing and controlling, leadership is most evident in the directing function.

Randolph and Posner (1988) provided these ten guidelines for effective project management, all very appropriate for design projects:

1. Set a clear project goal;
2. Determine the project objectives for team members;
3. Establish check-points, activities, relationships, and time estimates;
4. Draw a picture of the project schedule;
5. Direct people individually and as a project team;
6. Reinforce the commitment and excitement of the project team;
7. Keep everyone connected with the project informed;
8. Build agreements that vitalize team members;
9. Empower yourself and others on the project team; and
10. Encourage risk taking and creativity.

These ten guidelines, when implemented for managing design innovation will really help integrate the technical and human aspects of project management and will lead to completing the project on time, within budget, and according to the desired quality standards.

7.3.2 *Team responsibilities*

Although not everybody will become a leader or a manager, those who follow or are managed are of equal importance where team-work is concerned. Moreover, leadership stems from two sources: being like the people who are to be led, and – at the same time – being different. Like them, by being representative of their characteristics; different, by being better able to understand the problems, see workable solutions and make those solutions clear and acceptable to the followers. Therefore, there should be a close relationship between leaders and followers.

It is generally recognized that top design or research engineers are more important than the managers of the design projects on which they work. As a result, some leading corporations – IBM, GE, ITT for example – have a two tier system of administration and professional hierarchies. The old practice of promoting highly competent designers into better-paying managerial positions at which they may not be as good ('own level of incompetence,' by the 'Peter Principle') is gradually being abandoned. It is not uncommon that, at IBM, a design engineer will be paid more than the project manager, all compatible with the qualifications and experience of each. Other companies, such as DuPont and Gore, Inc. go even further. In contrast to the traditional pyramid, a system of management called 'lattice organization' was created. The lattice structure encourages project teams made up of various specialists *without specifically assigned leaders*. If the team runs into an engineering problem, a team engineer takes over; if it is a marketing problem, then the marketing specialist will be in charge. In this way, there are no bosses but there are always leaders.

The lattice structure is important because it gives the opportunity of leadership to anyone qualified and willing, and encourages each individual to deal directly with every one to get a job done, rather than work through a traditional chain of command. In essence, the system relies on self-management and voluntary commitments.

Consequently, anyone connected with design innovation, whether in administration or engineering must be familiar with the basic concepts of management and leadership. The old distinction of 'leaders' and 'followers' is outmoded: they are really co-innovators (Clark and Clark, 1990). Such a philosophy also leads to new awareness of management and leadership by engineers. Currently, two-thirds of the seats on the boards of American companies are occupied by people trained in law, finance, or accounting. Only 81 of the CEOs of the top 1,000 US companies were educated in engineering or science. By contrast, in Japan more than 65% of the board of directors of the leading companies graduated from engineering and science programs and not from graduate schools of business. In France, most of the leaders of business and government have graduated from the elite 175 Grand Ecoles of engineering and technology (e.g. Ecole Politechnique in Paris). In Germany, a majority of corporate leaders are alumni of the technical universities, whose graduate engineers have completed a period in industry and a thesis on an industrial problem.

An excellent paper specifically for engineering design managers, was prepared by Kramer (1987) and should be of value to American technology managers interested in the thinking of their European counterparts. Finally, for those wishing to improve their managerial skills or who still doubt the emergence and value of the 'management of design innovation' concept, compulsory reading should be the entire proceedings of the fairly recent International Conference on Design in Management of Business, held in Boston (Gregory, 1987).

7.4 WORKSHOP EXERCISES

The best way to learn the concepts discussed in this book is to participate in actual design projects and practice the various techniques for idea generation, innovative design, communication and management. During the author's workshops on these subjects, the participants – from research and industry – have come up with many inspiring problems and ingenious design solutions all of which were thought provoking as well as fun to work on.

In this section, a selection of the topics of the design exercises used at the workshops is listed. A solution for the conceptual design of each mini-project was developed in 2-3 days by design teams consisting of 3-4 persons. As an intellectual challenge to the reader, a series of other examples of design problems are also given. Solving these problems will enable the reader to demonstrate the newly acquired innovative design skills.

7.4.1 *Conceptual design mini-projects used at 2-3 day workshops*

1. An automated remote system for cutter head control of a continuous miner machine.
2. A temperature controlled garment for use in hot mines.
3. A robot system for advance reconnaissance and rescue after a mine fire.
4. Three-parameter roof classification system for selection of resin-grouted bolts in coal mines.
5. A device to perform remotely rockburst 'proof tests' on mine structures.
6. A high-extraction zero-subsidence mining method.
7. Oxygen supply system on the Moon, based on mining of oxygen-rich minerals from lunar soil.
8. A pre-sealed liner for a shaft in soft rock.
9. Remote and continuous profiling of the surface of a drill-and-blast tunnel.
10. An underground mining method for deep alluvial placer deposits.
11. A cost-effective shale sealant system.
12. A mining system for deep hard rock vein mines to replace drilling and blasting.
13. A physical see-thru model of groundwater interaction with backfill in an underground mine.
14. Mining of heavy oil by large hole drilling.
15. An electric motorcycle capable of one-week use and 300-mile range.

7.4.2 *General design exercises (after Suh, 1990)*

1. From the viewpoint of design principles, which is the better form of government, the presidential form (as in USA), or the parliamentary form (as in Europe).

2. Develop functional requirements and design components for a bicycle.

3. Design an educational system for providing lifelong learning to practicing engineers.

4. Consider the design of the human body. What are the functional requirements and design components? Is the human body a coupled system? Can you improve the design?

5. Design an ideal international trade system that satisfies at least the following functional requirements:

FR_1 = No country can have a surplus that is greater than 10% of its GNP.

FR_2 = No country can incur a foreign debt that requires annual interest payments of more than 10% of its GNP.

FR_3 = No country should have less per capita income than 5% of the average per capita income of OECD nations.

6. Mr. John Doe is looking for a house. He is considering two different towns, which are located near the town in which his company is located. His 'design range' in making his decision is based on three factors: (1) land price, (2) commuting time, and (3) environment. He is willing to pay up to $10,000 per acre, to spend as much as 1.5 hours for commuting, and to live in an environment which meets 80% of his expectations. The two towns selected offer the possibilities as listed below. Which is the better choice for Mr. Doe?

	Town A	Town B
Land price ($/acre)	4,000-9,000	9,200-12,000
Commuting time (hrs)	1.0-1.7	0.8-1.0
Environment (%)	80-90	60-85

Compendium of early design charts in rock engineering

These charts provide a historical perspective of the attempts to develop design methodologies in rock engineering during the past 25 years. They exemplify both interest in and the evolution of the design processes in mining, tunneling, slope and shaft engineering.

The charts that follow, in chronological order, have paved the way for the design methodology proposed in this book.

EARLIER DESIGN CHARTS FOR MINES, TUNNELS, SLOPES AND SHAFTS

Chart 1. Design of mine excavations, after Denkhaus (1968).

Chart 2. Planning a slope stability design program, after Hoek and Bray (1977).

Chart 3. Design of underground excavations in rock, after Hoek and Brown (1980).

Chart 4. Interaction of various factors in mine design, after Luxbacher and Ramani (1980).

Chart 5. Design guidelines for shaft design, after Roesner et al. (1983).

Chart 6. Detailed design procedure for rock tunnels, after Bieniawski (1984).

Chart 7. Design approach for strata control in mines, after Ramani and Prasad (1987).

Chart 8. Design process for tunneling according to the International Tunnelling Association (after Duddeck, 1988).

Chart 9. Design process steps for the Basalt Nuclear Waste Isolation Project, after Schmidt (1988).

Chart 10. Flow chart of proposed shaft liner design methodology for nuclear waste repositories, after Richardson (1988).

Chart 11. Recommended mine design procedure, after Lineberry and Adler (1989).

Chart 12. Logic chart for drift design methodology for nuclear waste repository in Tuff, after Hardy et al. (1990).

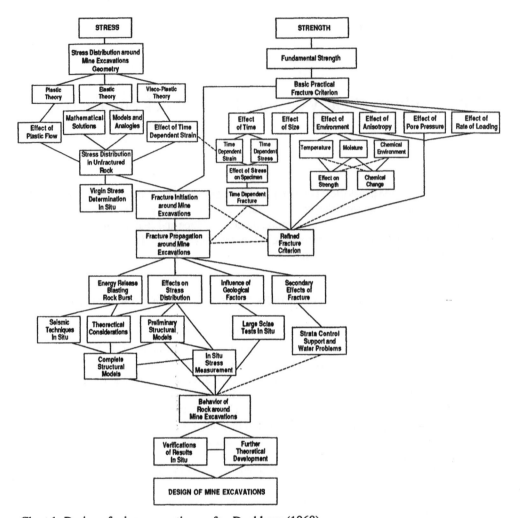

Chart 1. Design of mine excavations, after Denkhaus (1968).

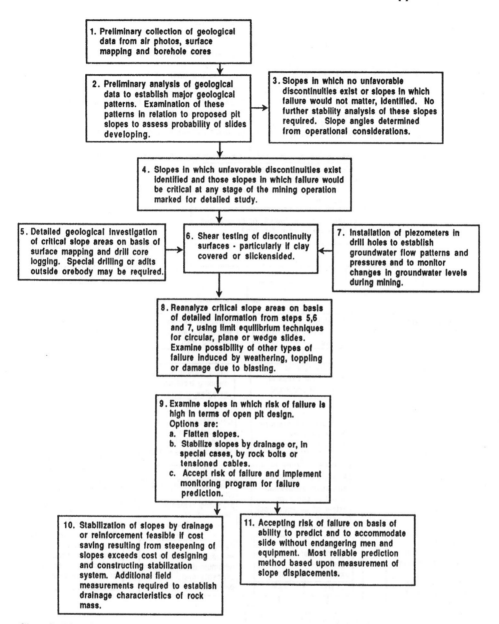

Chart 2. Planning a slope stability design program, after Hoek and Bray (1977).

Preliminary collection and interpretation of geological data from historical documents, geological maps, air photographs, surface mapping and borehole core logs. Consideration of the relationship between the rock mass characteristics and the geometry and orientation of the proposed excavations.

In hard rock masses with strongly developed inclined structural features, excavation stability may be dominated by gravity falls and sliding along inclined discontinuities. Rock classification systems inadequate.

When stability is not likely to be dominated by sliding on structural features, other factors such as high stress and weathering become important and can be evaluated by means of a classification of rock quality.

Use of rock quality index to compare excavation stability and support requirements with documented evidence from sites with similar geological conditions.

Are stability problems anticipated for excavations of size and shape under consideration?

| YES | NO |

Design of excavations based on operational considerations with provision for minimal support.

Instability due to adverse structural geology.

Instability due to excessively high rock stress.

Instability due to weathering and / or swelling rock.

Instability due to excessive groundwater pressure or flow.

Detailed geological mapping of borehole core, surface exposures, exploratory adits and shafts.

Measurement of in situ rock stress in vicinity of proposed excavations.

Slake durability and swelling tests on rock samples.

Installation of piezometers for determination of groundwater pressures and distribution.

Can stability be improved by relocation and/or reorientation of excavations?

| YES | NO |

Rock strength tests to determine rock fracture criterion.

Consideration of remedial measures such as pneumatically applied concrete lining.

Design of drainage and / or grouting system to control excessive groundwater pressure and flow into excavations.

Design of excavations with provision for close geological observation and local support as required.

Stress analysis of proposed excavation layout to check on extent of potential rock fracture.

Trial excavation to test effectiveness of proposed remedial measures.

Can rock fracture be minimized or eliminated by change of excavation layout?

| NO | YES |

Design of excavation sequence to ensure minimum delay between exposure and protection of surfaces.

Provision of permanent ground-water monitoring facilities to check continuing effectiveness of drainage measures.

Design of support to prevent gravity falls and to reinforce potential fracture zones.

Design of excavations with provision for trial excavation, controlled blasting, rapid support installation and monitoring of excavation behavior during and on completion of construction.

Can adequate support be provided to ensure long term stability?

| NO | YES |

Reject this site

Chart 3. Design of underground excavations in rock, after Hoek and Brown (1980).

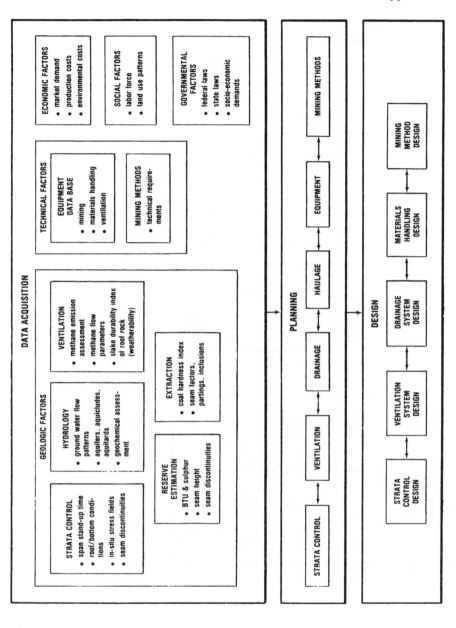

Chart 4. Interaction of various factors in mine design, after Luxbacher and Ramani (1980).

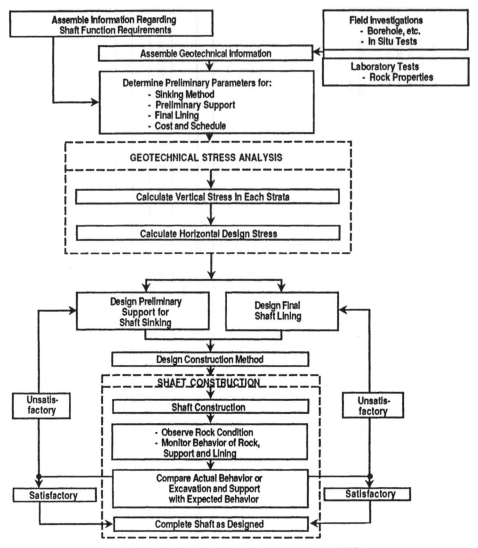

Chart 5. Design guidelines for shaft design, after Roesner et al. (1983).

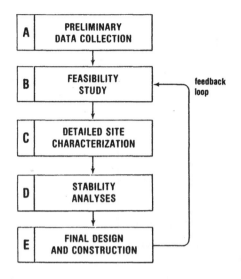

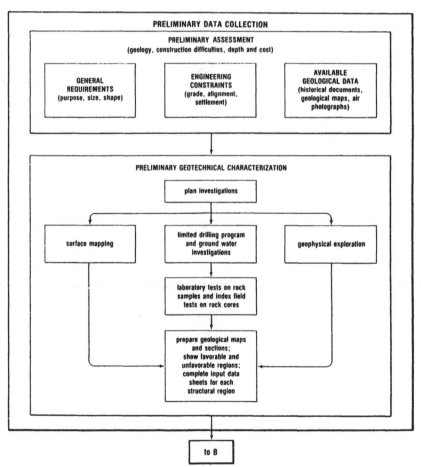

Chart 6A. Detailed design procedure for rock tunnels, after Bieniawski (1984).

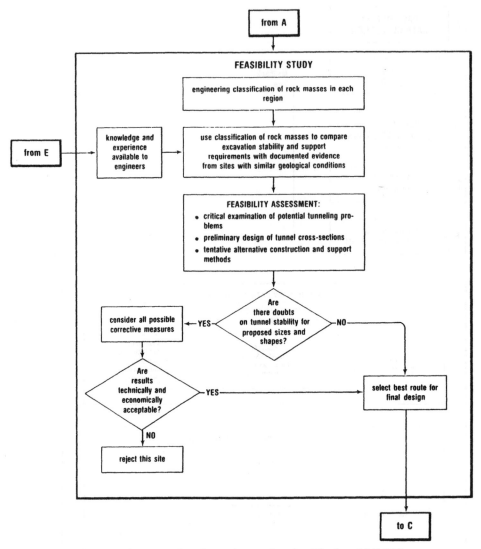

Chart 6B. Detailed design procedure for rock tunnels, after Bieniawski (1984).

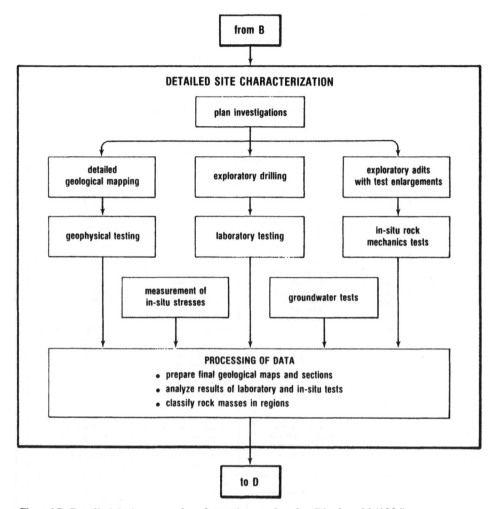

Chart 6C. Detailed design procedure for rock tunnels, after Bieniawski (1984).

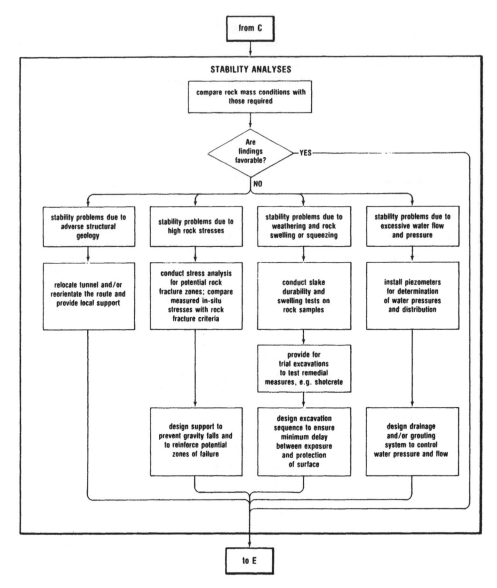

Chart 6D. Detailed design procedure for rock tunnels, after Bieniawski (1984).

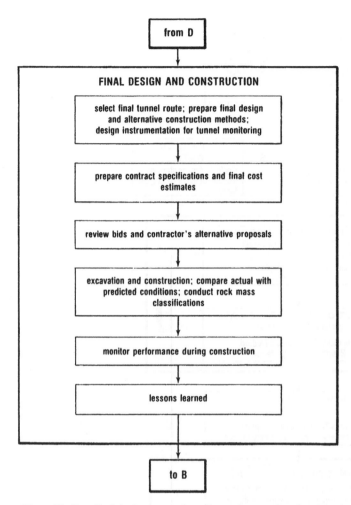

Chart 6E. Detailed design procedure for rock tunnels, after Bieniawski (1984).

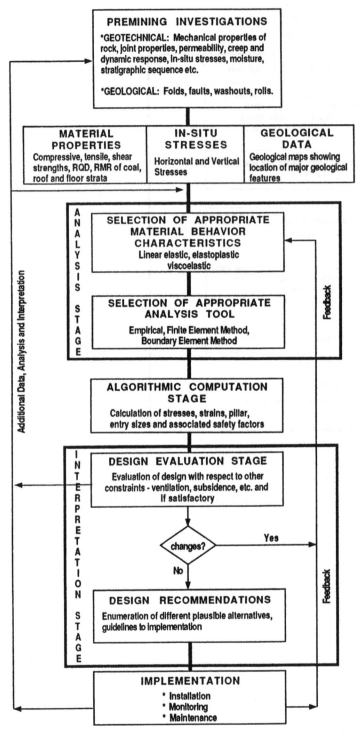

Chart 7. Design approach for strata control in mines, after Ramani and Prasad (1987).

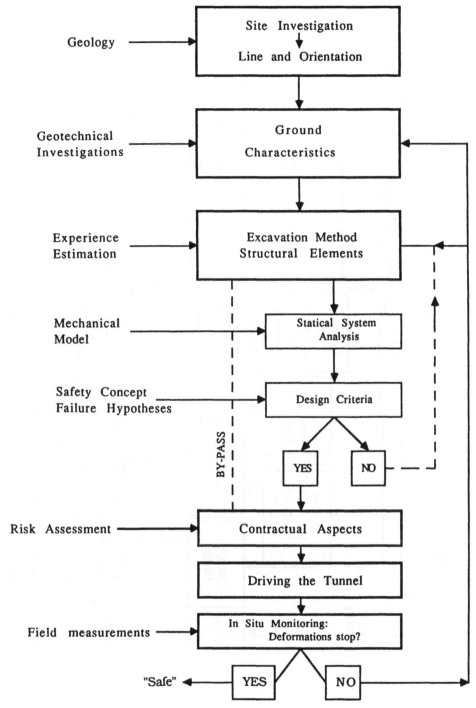

Chart 8. Design process for tunneling according to the International Tunnelling Association (after Duddeck, 1988).

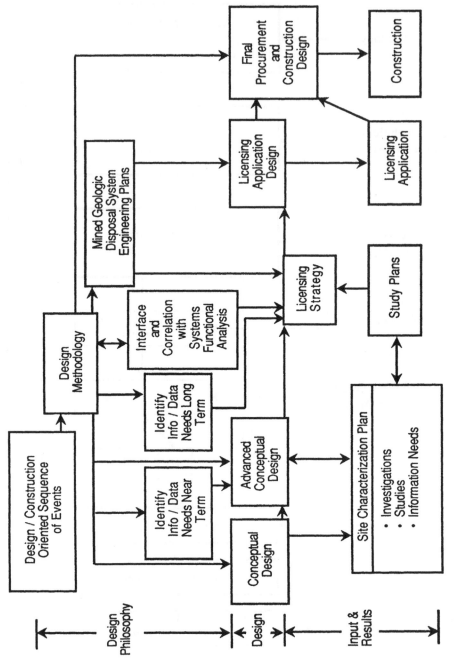

Chart 9. Design process steps for the Basalt Nuclear Waste Isolation Project, after Schmidt (1988).

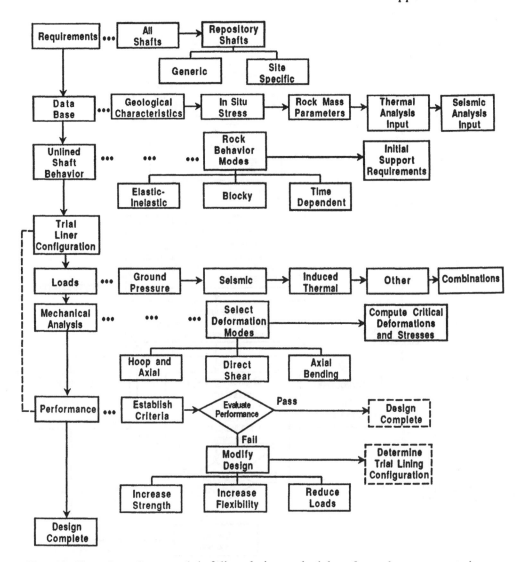

Chart 10. Flow chart of proposed shaft liner design methodology for nuclear waste repositories, after Richardson (1988).

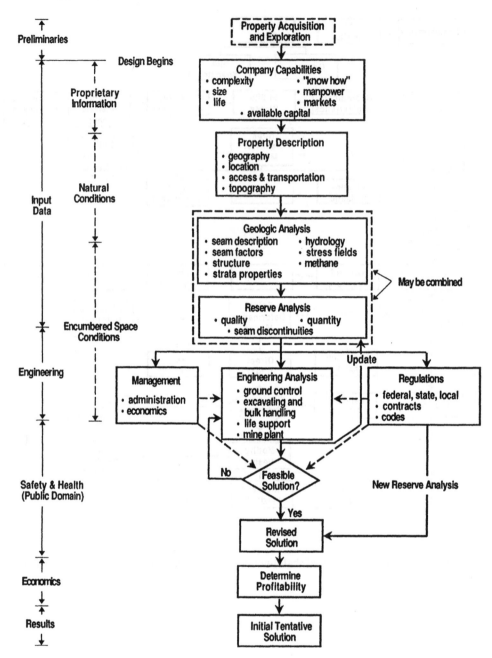

Chart 11. Recommended mine design procedure, after Lineberry and Adler (1989).

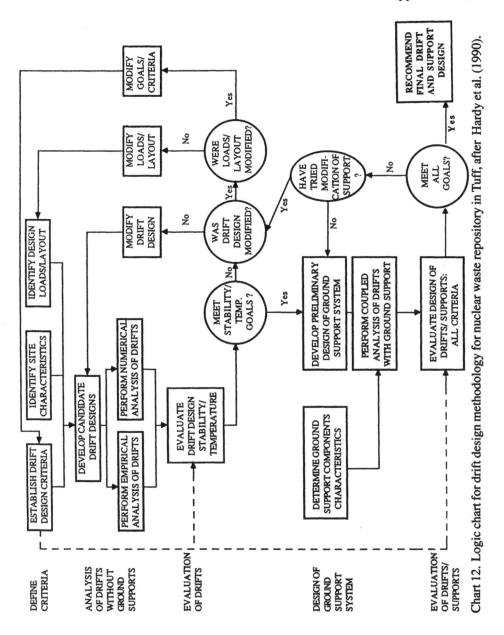

Chart 12. Logic chart for drift design methodology for nuclear waste repository in Tuff, after Hardy et al. (1990).

Bibliography

Accreditation Board for Engineering and Technology. *Fifth Annual Report*, 117 pp. ABET, Washington, DC (1987).

Adams, J.L. *Conceptual Blockbusting*, 161 pp. Addison-Wesley Publishing Co., New York (1988).

Adelson, B. Cognitive research: uncovering how designers design; cognitive modeling: explaining and predicting how designers design. *Research in Engineering Design* 1, pp. 35-42 (1989).

Alic, J.A. The rise and fall of R & D. *Mechanical Engineer* 110, pp. 28-33 (1988).

American Society for Engineering Education. *A National Action Agenda for Engineering Education*, 39 pp. ASEE, Washington, DC (1987).

American Society of Mechanical Engineers. *Goals and Priorities for Research in Engineering Design*, 77 pp. ASME, New York (1986).

Andrews, J.G. In search of the engineering method. *Engineering Education* 78, pp. 29-30, 55-57 (1987).

Archer, L.B. Systematic method for designers. In *Developments in Design Methodology* (Edited by N. Cross), pp. 55-82. John Wiley & Sons, New York (1984).

Archiszewski, T. Design theory and methodology in Eastern Europe. In *Design Theory and Methodology '90* (Edited by J.R. Rinderly), pp. 209-217. ASME, New York (1990).

Asimow, M. *Introduction to Engineering Design*, 135 pp. Prentice-Hall, Englewood Cliffs, New Jersey (1962).

Barr, V. Brainstorming and storyboarding. *Mechanical Engineering* 110, pp. 42-46 (1988).

Bazjanac, V. Architectural design theory: models of the design process. In *Basic Questions of Design Theory* (Edited by W.R. Spillers), pp. 3-19. North-Holland, Amsterdam (1974).

Beitz, W. General approach to systematic design – application of VDI Guideline 2221. In *Proceedings, 1987 International Conference on Engineering Design* (Edited by W.E. Eder), pp. 15-20. ASME, New York (1987).

Bennett, W.J. *James Madison High School – A Curriculum for American Students*, 49 pp. US Department of Education, Washington, DC (1987).

Bieniawski, Z.T. Elandsberg pumped storage scheme. In *Exploration for Rock Engineering* (Edited by Z.T. Bieniawski), pp. 272-289. Balkema, Rotterdam (1976).

Bieniawski, Z.T. A critical assessment of selected in-situ tests for rock mass deformability and stress measurements. In *Proc. 19th US Symp. on Rock Mechanics*, pp. 523-535. University of Nevada, Reno (1978).

Bieniawski Z.T. Determining rock mass deformability: Experience from case histories. *Int. J. Rock Mech. Min. Sci.*, 15, pp. 237-247 (1978).

Bieniawski, Z.T. The 'Petite Sismique' technique – a review of current developments. In *Proc. 2nd Conference on Acoustic Emission/Microseismic Activity in Geologic Structures and Materials*, pp. 305-318. Trans Tech Publications, Clausthal, Germany (1980).

Bieniawski, Z.T. The design process in rock engineering. *Rock Mech. and Rock Engineering* 17, pp. 183-190 (1984).

Bieniawski, Z.T. *Rock Mechanics Design in Mining and Tunneling*, 272 pp. Balkema, Rotterdam (1984).

Bieniawski, Z.T. The design of underground repositories for storage of high level nuclear waste. In *MEXROC'85: The Role of Rock Mechanics in Excavations for Mining and Civil Works* (Edited by R.S. Trejo), pp. 595-604. US National Committee on Rock Mechanics, Washington, DC (1985).

Bieniawski, Z.T. Principles of engineering design for strata control in mining. In *Applications of Rock Characterization Techniques in Mine Design* (Edited by M.Karmis), pp. 3-10. AIME, New York (1986).

Bieniawski, Z.T. *Strata Control in Mineral Engineering*, 212 pp. Balkema, Rotterdam/John Wiley and Sons, New York (1987).

Bieniawski, Z.T. Towards a creative design process in mining. *Mining Engineering* 40, pp. 1040-1044 (1988).

Bieniawski, Z.T. *Engineering Rock Mass Classifications*, 251 pp. John Wiley & Sons, New York (1989).

Brown, E.T. (Editor). *Rock Characterization, Testing and Monitoring – ISRM Suggested Methods*, 211 pp. Pergamon Press, Oxford (1981).

Brown E.T. From theory to practice in rock engineering. *Trans. Inst. Min. Metall.* 94, pp. A67-A83 (1985).

Bucciarelli, L.L. An ethnographic perspective on engineering design. *Design Studies* 9, pp. 159-168 (1988).

Bucciarelli, L.L., Goldschmidt, G. and Schon, D.A. Generic design process in architecture and engineering. In *Proceedings, 1987 Conference on Planning and Design in Architecture* (Edited by J.P. Protzen), pp. 59-64. ASME, New York (1987).

Chaplin, C.R. *Creativity in Engineering Design: the Educational Function*, 45 pp. Report no. FE4, The Fellowship of Engineering, London (1989).

Clark, K.E. and Clark, M.B. (Editors). *Measures of Leadership*, 636 pp. Center for Creative Leadership, Greensboro, NC (1990).

Cohen, R.W. The innovation process – a national problem. In *2nd International Conference on Creativity and Innovation* (Edited by S.S. Gryskiewicz), pp. 222-246. Center for Creative Leadership, Greensboro, NC (1988).

Crochetiere W.J. Engineering design education at Tufts University – the first twenty years. In *Proceedings, 1987 International Conference on Engineering Design* (Edited by W.E. Eder), pp. 1028-1032. ASME, New York (1987).

Cross, N. The nature of design activity. In *Development in Design Methodology* (Edited by N. Cross), pp. 167-173. John Wiley & Sons, New York (1984).

Cross, N. *Engineering Design Methods*, 159 pp. John Wiley & Sons, New York (1989).

de Bono, E. *The Use of Lateral Thinking*, 101 pp. Jonathan Cape, London (1967).

Denkhaus, H.G. Scientific development of rock mechanics during the past ten years: Conclusions from rock mechanics knowledge in reference to construction and stability of underground excavations. In *Bericht (Proceedings), 10. Ländertreffen des Internationalen Büros für Gebirgsmechanik*, pp. 72-80. Akademie-Verlag, Berlin (1968).

Devon, F.R. The elusive definition of the engineering method. *Engineering Education* 79, pp. 12-13 (1988).

Dietrych, J. The integral information system – problems of design policy. In *Proceedings, 1987 International Conference on Engineering Design* (Edited by W.E. Eder), pp. 21-28. ASME, New York (1987).

Dixon, J.R. The state of education. *Mechanical Engineering* 113 (2), pp. 64-91 (1991).

Dixon, J.R. New goals for engineering education. *Mechanical Engineering* 113 (3), pp. 56-62 (1991).

Dixon, J.R. and Finger, S. The proper study of mankind is the science of design. *Research in Engineering Design* 1, pp. 1-2 (1989).

Duddeck, H. (Editor). Guideline for the design of tunnels. *Tunnelling and Underground Space Technology* 3, pp. 237-249 (1988).

Eder, W.E. Engineering education: needs and recommendations for a design-based curriculum. In *Proceedings, International Conference on Engineering Design*, pp. 2117-2125. Yugoslav Society of Engineers, Dubrovnik (1990).

Eder, W.E. Theory of technical systems: prerequisite to design theory. In *Proceedings, 1987 International Conference on Engineering Design* (Edited by W.E. Eder), pp. 103-113. ASME, New York (1987).

Ehrlenspiel, K. and John, T. Inventing by design methodology. In *Proceedings, 1987 International Conference on Engineering Design* (Edited by W.E. Eder), pp. 29-37. ASME, New York (1987).

Ericsson, K. and Simon, H. *Protocol Analysis*, 111 pp. MIT Press, Cambridge, MA (1984).

Ernst, E.W. and Lohmann, J.R. Designing undergraduate design curricula. *Engineering Education* 80, pp. 543-547 (1990).

Evans, B., Powell, J.A. and Talbot, R. (Editors). *Changing Design*, 379 pp. John Wiley & Sons, New York (1982).

Evans, D.L., McNeill, B.W. and Beakley, G.C. Design in engineering education: past views of future directions. *Engineering Education* 80, pp. 517-522 (1990).

Fairhurst, C. The application of mechanics to rock engineering. In *Exploration for Rock Engineering* (Edited by Z.T. Bieniawski), pp. 1-22. Balkema, Rotterdam (1976).

Fairweather, V. Radioactive waste: finding a safe place. *Civil Engineering* 61, pp. 48-51 (1991).

Faria Santos, C. and Bieniawski, Z.T. Floor design in underground coal mines. *Rock Mechanics and Rock Engineering* 22, pp. 249-271 (1989).

Fettweis, G. B. Rock mechanics as part of a mining engineering sub-discipline: geo-mining conditions. *Mineral Resources Engineering* 2, pp. 213-223 (1989).

Fiedler, F. How to be a successful leader. *The Toastmaster*, October, pp. 11-15 (1980).

Finger, S. and Dixon, J.R. A review of research in mechanical engineering design. *Research in Engineering Design* 1, pp. 51-67 (1989).

Florman, S.C. *The Existential Pleasures of Engineering*, 130 pp. St. Martin's Press, New York (1976).

Fogue, R. Beyond design methods – arguments for a practical design theory. In *Changing Design* (Edited by B. Evans, J.A. Powell and R. Talbot), pp. 257-274. John Wiley & Sons, New York (1982).

Franklin J.A. and Dusseault M.B. *Rock Engineering*, 601 pp. McGraw-Hill, New York (1989).

Gasparski, W. Design methodology: a personal statement. In *Philosophy of Technology* (Edited by P.T. Durbin), pp. 153-167. Kluwer Academic Publishers, Amsterdam (1989).

Gentry, D.W. *Minerals curricula in transition*, 27 pp. Annual Meeting, National Association of Mineral Institute Directors, Washington, DC (1990).

Gergowicz, Z. Rock mechanics as a substantial thread in the educational process of mining engineers. *Mineral Resources Engineering*, 3, pp. 3-4 (1990).

Gill, H. The nature of problems. In *Proceedings, 1987 International Conference on Engineering Design* (Edited by W.E. Eder), pp. 114-122. ASME, New York (1987).

Glegg, G.L. *The Design of Design*, 93 pp. Cambridge University Press, Cambridge (1969).

Gregory, S.A. *Design Sciences*, 323 pp. Butterworths, London (1986).

Gregory, S.A. Creativity for the 1990s in design and management. In *Proceedings, 1987 Conference on Planning and Design in Management of Business and Organization* (Edited by P.C. Nutt), pp. 61-64. ASME, New York (1987).

Gryskiewicz, S.S. Creative problem solving. *Planned Innovation* 3, pp. 3-5 (1980).

Guralnik, D.B. (Editor) *Webster's New World Dictionary of the American Language*, 1275 pp. Simon and Schuster, New York (1986).

Hales, C. *Analysis of the Engineering Design Process in an Industrial Context*, 25 pp. Grants Hill Publications, Cambridge (1987).

Haley, A. Producing the Next Generation of Design Managers. London Business School, London, England. Personal communication (1989).

Hardy, M.P., Brechtel, C.E., Goodrich, R.R. and Bauer, S.J. Preliminary drift design analyses for nuclear waste repository in tuff. In *Proc. 31st U.S. Symp. on Rock Mechanics*, pp. 345-352. Balkema, Rotterdam (1990).

Harrisberger, L. *Engineersmanship – the Doing of Engineering Design*, 43 pp. Wadsworth Inc., Belmont, CA (1982).

Hazelrigg, G.A. In continuing search of the engineering method. *Engineering Education* 79, pp. 118-121 (1988).

Hickling, A. Beyond a linear iterative process? In *Changing Design* (Edited by B. Evans, J.A. Powell and R. Talbot), pp. 275-294. John Wiley & Sons, New York (1982).

Hoek, E. Practical rock mechanics – developments over the past 25 years. *Trans. Instn Min. Metall.* 96, pp. A1-A-6 (1987).

Hoek, E. and Bray, J.W. *Rock Slope Engineering*, pp. 13-16. Institution of Mining and Metallurgy, London (1977).

Hoek, E. and Brown, E.T. *Underground Excavations in Rock*, pp. 9-12. Institution of Mining and Metallurgy, London (1980).

Huang, Z.J. The epochal significance of design methodology and our tasks. In *Proceedings, 1987 International Conference on Engineering Design* (Edited by W.E. Eder), pp. 46-51. ASME, New York (1987).

Hubka, V. *Principles of Engineering Design*, 118 pp. Heurista/Springer-Verlag, New York (1987).

Hubka, V. and Eder, W.E. *Theory of Technical Systems*, 275 pp. Springer-Verlag, New York (1988).

International Society for Rock Mechanics. Report on the teaching of rock mechanics. *Int. J. Rock Mech. Min. Sci.* 20, pp. 189-200 (1983).

Isaksen, S.G. Toward a model for the facilitation of creative problem solving. *Journal of Creative Behavior* 17, pp. 18-31 (1987).

Jansson, D.G. Design fixation. In *Proceedings, NSF Engineering Design Research Conference*, pp. 54-76. University of Massachusetts, Amherst (1989).

Jones, C.J. A method of systematic design. In *Conference on Design Methods* (Edited by C.J. Jones and D.G. Thornley), pp. 53-73. Macmillan, New York (1963).

Jones, C.J. and Thornley, D.G. (Editors). *Conference on Design Methods*, 222 pp. MacMillan, New York (1963).

Kerley, J.J. *Creative Inventive Design and Research*, 133 pp. NASA Goddard Space Flight Center, Greenbelt, MA (1986).

Kicker, D.C. The Theory and Methodology of Rock Mechanics Design, 213 pp. Ph.D. Thesis, Pennsylvania State University, University Park, PA (1990).

Klein, H.E. (Editor). *Problem Solving with Cases and Simulations*, 546 pp. World Association for Case Method Research and Application, Boston, MA (1990).

Klir, G.J. *An Approach to General Systems Theory*, 123 pp. Van Nostrand, New York (1969).

Koen, B.V. Toward a definition of the engineering method. *Engineering Education* 75, pp. 150-155 (1984).

Koen, B.V. *Design of the Engineering Method*, 75 pp. American Society for Engineering Education, Washington, DC (1985).

Koen, B.V. The engineering method and the state-of-the-art. *Engineering Education* 77, pp. 670-674 (1986)

Kotarbinski, T. *Praxiology: Introduction to the Science of Efficient Action*, 176 pp. Pergamon Press Oxford (1965).

Kramer, F. Strategic technology management as a basis for successful design innovation. *Konstruktion* 39, pp. 259-266 (1987).

Landau, R. Methodology of research programs. In *Changing Design* (Edited by B. Evans, J.A. Powell and R. Talbot), pp. 303-310. John Wiley & Sons, New York (1982).

Langer, M. Waste disposal in the Federal Republic of Germany. *Bulletin of Int. Association of Engineering Geology* 39, pp. 53-58 (1989).

Lewis, L.S. and Kingston, P.W. The best, the Brightest and the most affluent. *Academia*, December, pp. 28-33 (1989).

Lineberry, G.T. and Adler, L. A model for underground coal mine design. In *Proceedings, Multinational Conference on Mine Planning and Design*, pp. 59-69. University of Kentucky, Lexington, KT (1989).

Loire, R. *The Design Way*, 2002 pp. Tramco, Paris/A. Ghosh, Houston (1989).

Luxbacher, W.G. and Ramani, R.V. The interrelationship between coal mine plant and ventilation system design. In *Proceedings, 2nd International Mine Ventilation Congress*, pp. 73-82. AIME, New York (1980).

Mark, H. The fourth revolution: educating engineers for leadership. *Engineering Education* 79, pp. 104-108 (1988).

Mark, C. and Bieniawski, Z.T. An empirical method for the design of chain pillars in longwall mining. In *Proc. 27th U.S. Symp. on Rock Mechanics*, pp. 415-422. AIME, New York (1986).

McAleer, N. Creativity: the roots of inspiration. *Omni* 11, pp. 42-48 (1989).

McCall, R.J. Issue-serve systems: a descriptive theory of design. *Design Methods and Theories* 20, pp. 443-458 (1986).

McMahon, E.H. Designing the engineering graduate. *Bulletin of Design in Engineering Education*, 14, pp. 14-19 (1989).

McMasters, J.H. and Ford, S.D. An industry view of enhancing design education. *Engineering Education* 80, pp. 526-539 (1990).

McNeill, B.W., Evans, D.L., Bellamy, L. and Beakley, G.C. Beginning design education with freshmen. *Engineering Education* 80, pp. 548-553 (1990).

Miller, W.C. The creative edge: fostering innovation where you work. *Design News* 3, pp. 404-405 (1989).

Mitchell, W.J. *The Logic of Architecture: Design, Computation and Cognition*, 292 pp. MIT Press, Cambridge, MA (1990).

Montalbo, T. Flex your mental muscles. *The Toastmaster*, August, pp. 8-10 (1988).

Morfeldt, C.O. Different subsurface facilities for geologic disposal of radioactive waste in Sweden. *Bulletin of Int. Association of Engineering Geology* no. 39, pp. 25-33 (1989).

Nadler, G. Systems methodology and design. *Mechanical Engineering* 108, pp. 84-88 (1986).

Nadler, G. and Hibino, S. *Breakthrough Thinking*, 350 pp. Prima Publishing & Communications, Rocklin, CA (1990).

National Academy of Sciences (Board on Radioactive Waste Management). *Rethinking High-Level Radioactive Waste Disposal*, 38 pp. National Academy Press, Washington, DC (1990).

National Research Council. *Competitiveness of the U.S. Minerals and Metals Industry*, 140 pp. National Academy Press, Washington, DC (1990).

National Research Council. *Improving Engineering Design*, 107 pp. National Academy Press, Washington, DC (1991).

Nicholson, G.A. and Bieniawski, Z.T. A nonlinear deformation modulus based on rock mass classification. *Int. J. of Mining and Geological Engineering* 8, pp. 181-202 (1990).

Nuclear Waste Technical Review Board. *Second Report to U.S. Congress and the U.S. Secretary of Energy*, 34 pp. Washington, DC (1990).

Olds, B.M., Pavelich, M.J. and Yeats, F.R. Teaching the design process to freshmen and sophomores. *Engineering Education* 80, pp. 554-559 (1990).

Owen, C.L. Design education in the information age. *Design Studies* 11, pp. 202-206 (1990).

Pahl, G. and Beitz, W. *Engineering Design*, 450 pp. Springer-Verlag, New York (1984).

Petroski, H. Engineering on the backs of envelopes. *American Scientist* 79, pp. 15-17 (1991).

Pugh, S. Design Courses for Managers at the University of Strathclyde, Glasgow, Scotland. Personal communication (1989).

Rabins, M. (Editor). Design theory and methodology – a new discipline. *Mechanical Engineering* 108, pp. 23-27 (1986).

Rallis, C.J. Design – the goal of engineering activity. *S. Afr. Mechanical Engineer* 22, pp. 62-71 (1963).

Ramani, R.V. and Prasad, K.V.K. Applications of knowledge based systems in mining engineering. In *Proceedings, 20th International Symposium on Application of Computers in Mining*, pp. 167-180. SAIMM, Johannesburg (1987).

Randolph, W.A. and Posner, B.Z. What every manager needs to know about project management. *MIT Sloan Management Review* 29, pp. 65-73 (1988).

Rauhut, F.J. The German viewpoint of the aims for the training of young mining engineers. *Mineral Resources Engineering* 2, pp. 203-211 (1989).

Richardson, A.M and St. John, C. A concrete shaft liner design methodology for nuclear waste repositories. In *Proc. 29th U.S. Symp. on Rock Mechanics*, pp. 653-664. Balkema, Rotterdam (1988).

Rittel, H.W.J. Second-generation design methods. In *Developments in Design Methodology* (Edited by N.Cross), pp. 317-328. John Wiley & Sons, New York (1984).

Rittel, H.W.J. and Weber, M.M. Planning problems are wicked problems. In *Developments in Design Methodology* (Edited by N. Cross), pp. 135-144. John Wiley & Sons, New York (1984).

Rodenacker, W.G. *Methodisches Konstruieren*. Springer-Verlag, Berlin (1970).

Roesner, E.K., Poppen, S.A.G. and Konopka, J.C. Stability during shaft sinking – a design guideline for ground support of circular shafts. In *Proceedings, 1st International Conference on Stability in Underground Mining*, pp. 749-769. AIME, New York (1983).

Rowe, P.G. *Design Thinking*, 229 pp. MIT Press, Cambridge (1987).

Queijo, J. Inside the creative mind. *Bostonia* 61, pp. 26-38 (1987).

Salamon, M.D.G. Developments in rock mechanics: a perspective of 25 years. *Trans. Instn. Min. Metall.* 97, pp. A57-A68 (1988).

Sanders, M.S. and Peay, J.M. *Human Factors in Mining*, 146 pp. Bureau of Mines Information Circular 9182, US Department on the Interior, Washington, DC (1988).

Sauer, G. When an invention is something new: from practice to theory of tunnelling. *Tunnels & Tunnelling* 20, pp. 35-39 (1988).

Schmidt, B. Learning from nuclear waste repository design: the ground-control plan. *Tunnelling and Underground Space Technology* 3, pp. 175-181 (1988).

Shaw, T. Education of mining engineers. *Mineral Resources Engineering* 2, pp. 201-202 (1989).

Siddall J.N. *Optimal Engineering Design*, 536 pp. Marcel Dekker Inc., New York (1982).

Simon, H.A. *The Science of the Artificial*, 91 pp. MIT Press, Cambridge, MA (1969).

Simon, H.A. What we know about the creative process. In *Frontiers in Creative and Innovative Management* (Edited by R.L. Kuhn), pp. 100-111. Ballinger Publishing Co., Cambridge, MA (1985).

Smalley, B. Creativity at the work place. *US Air Magazine*, March, pp. 31-36 (1986).

Smith, E.T. Are you creative? *Business Week*, September 30, pp. 80-84 (1985).

Starfield, A.M. and Cundall, P.A. Towards a methodology for rock mechanics modelling. *Int. J. Rock Mech. Min. Sci.* 25, pp. 99-106 (1989).

Stauffer, L.A. and Ullman, D.G. A comparison of the results of empirical studies into the mechanical design process. *Design Studies* 9, pp. 107-114 (1988).

Stauffer, L.A., Ullman, D.G. and Dietterich, T.G. Protocol analysis of mechanical engineering design. In *Proceedings, 1987 International Conference on Engineering Design* (Edited by W.E. Eder), pp. 74-85. ASME, New York (1987).

Steinbreder, H.J. The innovators. *Fortune Magazine*, June 6, pp. 51-64 (1988).

Stevens, A.L. and Costin, L.S. *Findings of the ESF Alternatives Study,* 68 pp. Sandia National Laboratories, Albuquerque, NM, report no. SAND90-3232 (1991).

Strzalecki, A. Styles of solving design problems: Notes from research into design creativity. *Design Methods and Theories* 15, pp. 127-136 (1982).

Suh N.P. *The Principles of Design*, 401 pp. Oxford University Press, New York (1990).

Tarricone, P. Education at a crossroads. *Civil Engineering* 60, pp. 74-77 (1990).

Technion (Israel Institute of Technology). Engineering education 2001. *Engineering Education* 78, pp. 105-124 (1987).

Tomiyama, T. Engineering design research in Japan. In *Design Theory and Methodology '90* (Edited by J.R. Rinderly), pp. 219-223. ASME, New York (1990).

Tomiyama, T. and Yoshikawa, H. Extended general design theory. In *Design Theory for CAD* (Edited by H. Yoshikawa and E.A. Warman), pp. 95-130. North-Holland, Amsterdam (1987).

Ullman, D.G. and Dietterich, T.G. Mechanical design methodology. In *Computers in Engineering*, pp. 173-180. ASME, New York (1986).

US National Committee on Tunneling Technology. *Geotechnical Site Investigations for Underground Projects*, 182 pp. National Academy Press, Washington, DC (1984).

VDI Richtlinie 2221: *Systematic Approach to the Design of Technical Systems and Products*, 34 pp. Verein Deutscher Ingeniere, Düsseldorf (1987).

Viets, H. Designing across the curriculum. *Engineering Education* 80, pp. 565-567 (1990).

Wallace, K.M. and Hales, C. Some applications of a systematic design approach in Britain *Konstruktion* 39, pp. 275-279 (1987).

Weisberg, R.W. *Creativity: Genius and other Myths*, 222 pp. Freeman, New York (1986).

Whitney, D.E. Designing the design process. *Research in Engineering Design* 2, pp. 3-13 (1990).

Wilde, D.J. *Globally Optimal Design*, 288 pp. John Wiley & Sons, New York (1978).

Willem, R.A. Design and science. *Design Studies* 11, pp. 43-47 (1990).

Williamson, A. and Hudspeth, B. *Teaching Holistic Thought in Engineering Design*, 123 pp. Oregon State University Press, Corvallis, OR (1978).

Wilson, A.H. The stability of underground workings in the soft rocks of the coal measures. *Int. Journal of Mining Engineering* 1, pp. 91-187 (1983).

Wilson, W., Paterson, T. and Davies, A. Storing nuclear waste underground. *Tunnels and Tunnelling* 23, pp. 42-43 (1991).

Wrapp, H.E. Good managers don't make policy decisions. *Harvard Business Review* 64, pp. 8-16 (1984).

Yoshikawa, H. General design theory and a CAD system. In *Man – Machine Communication in CAD/CAM* (Edited by T. Sata and E.A. Warman), pp. 35-58. North-Holland, Amsterdam (1981).

Index

197

T - #0041 - 101024 - C1 - 254/178/11 [13] - CB - 9789054101260 - Gloss Lamination